Series:
Current Plant Science: Novel Concepts and Innovative Strategies

Genomic Selection in Plants
A Guide for Breeders

Editors

Ani A. Elias
Ramalingaswami Fellow
Institute of Forest Genetics and Tree Breeding
Coimbatore, Tamil Nadu, India

Shailendra Goel
Professor, Department of Botany
University of Delhi, Delhi, India

CRC Press
Taylor & Francis Group
Boca Raton London New York

CRC Press is an imprint of the
Taylor & Francis Group, an **informa** business
A SCIENCE PUBLISHERS BOOK

First edition published 2022
by CRC Press
6000 Broken Sound Parkway NW, Suite 300, Boca Raton, FL 33487-2742

and by CRC Press
4 Park Square, Milton Park, Abingdon, Oxon, OX14 4RN

CRC Press is an imprint of Taylor & Francis Group, LLC

Library of Congress Cataloging-in-Publication Data (applied for)

ISBN: 978-1-032-10350-1 (hbk)
ISBN: 978-1-032-10369-3 (pbk)
ISBN: 978-1-003-21499-1 (ebk)

DOI: 10.1201/9781003214991

Typeset in Times New Roman
by Radiant Productions

Preface

"Prediction of Total Genetic Value Using Genome-Wide Dense Marker Maps."
Meuwissen, T. H., Hayes, B. J. and Goddard, M. E. 2001.
Genetics 157(4): 1819–29.

That was how it all began! Around that time, plant as well as animal breeders were relying on phenotypic records, pedigree-based methods, and few identified genetic markers associated with a trait for the selection of individuals for economically important quantitative traits. With a great insight, Meuwissen and colleagues argued that all the markers in a genome contribute to the phenotypic variation of a quantitative trait and the use of such information would lead to faster genetic gain in breeding programs. This concept, now known as 'genomic selection', has enabled the breeders to select the genes for increased genetic gain within a short period of time in a cost-effective way. The concept of genomic selection, however, remained as a proposal until 2005 due to the lack of whole genome information as well as the presence of markers spread across the genome, something which changed with the advent of genome sequencing and gained speed with the revolution brought by Next Generation Sequencing (NGS).

The human genome project brought in the possibilities to identify diversity at the nucleotide level bringing in the concept of single nucleotide polymorphism (SNP), which is the most abundant form of genetic variation (Wang, 1998). Plant genome sequencing was lagging behind and the first eukaryotic plant genome of *Arabidopsis thaliana* was published only in year 2000 (The Arabidopsis Genome Initiative, 2000). It took another five years to publish the first genome sequence of an agriculturally important crop, rice (International Rice Genome Sequencing Project and Sasaki, 2005). The rice genome sequencing had a profound and immediate impact on rice improvement programs. Thanks to the NGS technologies such as Roche 454 Life Sciences, Illumina HiSeq and MiSeq, Sequencing by Oligonucloetide Ligation and Detection (SOLiD), years following 2005 saw publication of many plant genomes. Many of these technologies have now been replaced by more advanced methods of sequencing which brought down the cost. Long strand sequencing methods such as PacBio, Oxford Nanopore and 10X Genomics paved a way forward for sequencing more complex genomes rich in repetitive sequences. Further developments in NGS tools have provided methods of genotyping such as Genotyping by Sequencing (GBS) which placed SNPs as the marker of choice for high throughput (HTP) genotyping. This method has also made it possible to identify SNPs in polyploid species such as wheat and potato. Use of haplotype information together with allelic frequency helps in reducing false positives in SNPs identification in polyploids as reviewed by Mammadov et al. (2012).

With the availability of whole genome information and SNP markers, researchers including but not limited to Meuwissen, Hayes, Goddard, and vanRaden continued refining the genomic selection methods and models. While writing this book, 51,100 publications since 2005 are available on Google Scholar search using key words 'genomic selection' and 'plant breeding'. These publications provide access to methods, tools, case studies, datasets, as well as reviews facilitating the continuation of research progress in genomic selection (GS). Large international efforts in CIMMYT, ICRISAT, and Cornell University assured fruits of GS to the end user community.

In this book, we provide a comprehensive guide to plant breeders to implement GS in their crop improvement programs. The book starts by providing a general intuition on GS models and suggestions on how to choose an appropriate model for a crop based on the resources and the requirements. One of the factors that a breeder can control for increased prediction accuracy of the GS model is the design of the training population. Training population is used for developing a GS model which can later be tested and validated for prediction accuracy. Therefore, the second chapter is about the design of training population in the form of scripts. Breeders can directly use these scripts in open-source statistical platforms such as R. Extensive studies in cereal crops such as wheat, rice, and pearl millet; vegetables such as Solanaceae with a focus on tomato; legume crops chickpea and soybean; oil seed crops including groundnut, sunflower, oil palm, brassica, and other minor oil crops; and root crop potato are reviewed in the following chapters. The book ends with an insight to the advanced methods in GS for increasing genetic gain from the crop breeding programs.

Breeding is a complex endeavor considering the evolutionary force in nature. No method is a replacement for another; but the authors provide insight to multiple tools giving suggestions about design, management, and decision making in a breeding pipeline. In summary, we encourage breeders to perceive

GS as a tool with software, script language, and demonstrations available in theory and practice. We have provided open-source scripts, information on 'click and go' tools, links to genetic information, and theoretical explanation on GS.

Ani A. Elias
Shailendra Goel

References

International Rice Genome Sequencing Project and Takuji Sasaki. 2005. The map-based sequence of the rice genome. *Nature* 436(7052): 793–800. https://doi.org/10.1038/nature03895.

Mammadov, Jafar, Rajat Aggarwal, Ramesh Buyyarapu, Siva Kumpatla et al. 2012. SNP markers and their impact on plant breeding. *International Journal of Plant Genomics* 2012(December): 1–11. https://doi.org/10.1155/2012/728398.

The Arabidopsis Genome Initiative. 2000. Analysis of the genome sequence of the flowering plant *Arabidopsis thaliana*. *Nature* 408(6814): 796–815. https://doi.org/10.1038/35048692.

Wang, D. G. 1998. Large-scale identification, mapping, and genotyping of single-nucleotide polymorphisms in the human genome. *Science* 280(5366): 1077–82. https://doi.org/10.1126/science.280.5366.1077.

Contents

1

Genomic Selection Models for Plant Breeding Schemes: The Power of Choice

Ani A Elias

ABSTRACT

This chapter introduces key concepts of genomic prediction (GP) modeling and genomic selection (GS) approaches. The text is intended to serve as an in-depth introduction to GS for plant breeders. The chapter focuses on current models used in continuous traits, starting with an introduction to modeling approaches followed by a discussion on practical requirements for implementation. After the generalized introduction to the GS model and genomic relationship matrix, popular models in frequentist approach including panelized methods such as Genomic Best Linear Unbiased Prediction (GBLUP), Ridge Regression (RR), Least Absolute Shrinkage and Selection Operator (LASSO), and Elastic Net and kernel based methods such as Reproducing Kernel Hilbert Space (RKHS), and models in Bayesian main alphabet, their deviations and extensions are discussed. The chapter provides an intuition to the statistical models used in GS and GP aiming to provide suggestions while choosing a model for evaluation.

Institute of Forest Genetics and Tree Breeding, Coimbatore, Tamil Nadu, India - 641002.
Email: anianna01@gmail.com

Background

Genomic selection (GS) accelerates breeding pipelines by taking advantage of genomic information to generate genomic predictions (GP) of future performance without extensive phenotyping of selection candidates. The GP are generated by using a dataset that has both phenotypic and genomic information (a training set) to fix models that associate variations in genomic sequence or markers with observed phenotypic variation. The use of GP in GS breeding strategies can take many forms depending on the specific needs and resources of a given breeding program. Marker assisted selection (MAS) has been the method of choice for breeders to by-pass the extensive phenotyping. MAS, however, fails to capture vital minor gene effects as the method uses only major genes for the selection schemes (Shamshad and Sharma, 2018). In MAS, individual lines are selected based on the quantitative trait loci (QTL) detected through a significance test. Therefore, fewer markers are used for selection, ignoring genome wide variation for QTL with small to moderate effects. While small individual QTL may not contribute significantly to the observed phenotypes, the sum of all small QTL throughout the genome can explain a large portion of variation for complex traits. In GS, significance test is not used but the estimates are arranged for selection based on the accuracy in prediction and thus not based on the uncertainties in estimation (Bernardo, 2008; Meuwissen et al., 2001; Jannink et al., 2010; Arruda et al., 2016; Villar-Hernández et al., 2018).

GS removes the need for mapping genes, unlike in MAS where mapping is used to detect significant marker-QTL linkages in individuals. In GS, the markers are used directly in the model and/or used to calculate a genomic relationship matrix which is later used in the model. Using the markers, especially single nucleotide polymorphism (SNP), the alleles become the unit of analysis in GS (Knapp and Bridges, 1990). Alleles are the units to be replicated in the breeding scheme and they are replicated in the genotypes based on the genetic relatedness among the genotypes. This fact opens the possibility of testing more genotypes for the selection process thus increasing the selection intensity with more variation. Advances in high-throughput genotyping have markedly increased the viability of GS by driving down the cost of generating genome wide polymorphism information on selected candidates, making GS a cost-effective breeding strategy in breeding programs of all sizes across a wide range of crop species (Shamshad and Sharma, 2018).

In GS, the allelic substitution effect of SNP is estimated from a training population whose genotypic and phenotypic information is known. Based on this analysis, the genomic estimated breeding value (GEBV) is calculated for a test population by summing the SNP substitution effects given the SNP genotype information on a given test line.

$$GEBV_g = \sum_{j=1}^{J} x_{gj} * \beta_j \qquad [1]$$

where $GEBV_g$ is the genomic breeding value of individual g; x_{gj} is the gth genotype at the locus j (typically coded as a dosage of some reference allele); and β_j is the effect of the locus j.

Equation 1 represents a simple model that is complicated by the fact that the loci affecting the trait of interest are rarely known and the number of QTL affecting a trait is often very large relative to the number of phenotypic observations collected to train prediction models. This has profound implications on the types of modeling approaches used for GP and the optimal design of datasets used to train prediction models. Given the markers used for prediction do not necessarily represent the QTL controlling the trait of interest, prediction models rely on the non-random association of markers selected at random from the genome and the QTL associations. This non-random association is referred to as linkage disequilibrium (LD), which is the result of close physical linkage, limiting recombination, or non-random mating within the population. In scenarios where the trait of interest is not the trait that the QTL associated with the marker can best explain, consider changing the trait of interest or consider more than one trait of interest to begin with. An example from cattle milk yield is that polymorphism near gene SKC37A1 explains only 0.001% of the variance of the milk yield but 0.1% of the milk phosphorous (Kemper et al., 2016).

GEBV tend to have higher prediction accuracy compared to the BV estimated from the pedigree based relationship matrix (Habier et al., 2007) depending on the genetic architecture of the trait (Daetwyler et al., 2010), SNP marker density (Meuwissen et al., 2001; Meuwissen, 2009; Solberg et al., 2008), size of the training data (Meuwissen, 2009; Hoffstetter et al., 2016), genetic relatedness between the training and test populations (Habier et al., 2007), and population structures (Lyra et al., 2018). All these are factors that will influence LD between the markers used in the prediction model and the QTL controlling the trait of interest, in both the training and prediction datasets.

The genetic architecture of the trait refers to the number of QTL, the allelic effect on the phenotype, and the distribution of their effect (Heffner et al., 2011; Hill, 2012). Genome coverage is considered optimal when every QTL is in complete LD with at least one marker (Heffner et al., 2011). The optimal marker density depends on the effective population size (N_e) and genetic architecture of the trait. It is possible to achieve sufficient marker coverage in breeding programs using high density SNP platforms when genotyping by sequencing (GBS) is affordable (Heffner et al., 2011). Alternative genotyping strategies can leverage population and family structures to impute from lower density marker platforms to high density platforms for prediction across families to achieve optimal coverage at lower operation costs. Just for comparison, SNP markers which are biallelic require a density of two to three times that of SSR markers which are multiallelic. With a low marker density, markers are unlikely to be tightly linked to QTL and each marker may track a signal from different QTL, inducing a less extreme distribution of effects than obtained at higher density (Meuwissen et al., 2009). Furthermore, when LD patterns are not consistent between training and prediction sets, prediction accuracy would be expected to be low as the markers would be less likely to track the same QTL alleles in both datasets.

N_e is the number of individuals in an idealized population (random mating) that would result in the same rate of genetic drift (changes in allele frequency from one generation to the next) as the observed populations (Falconer and Mackay, 2009). N_e is an important concept in quantitative genetics which affects, among other things, the expected rate of LD decay. As such, N_e determines the number of independent chromosomal segments (M_e) segregating in a given population, where different independent segments trace back to different common ancestors (Hayes et al., 2009). Populations with large N_e require a large number of markers to achieve optimal coverage (e.g., Wray et al., 2007). Conversely, when N_e is sufficiently small optimal LD between SNP and potential QTL can be achieved using a smaller numbers of markers, and increasing LD using a denser panel is found not to have much effect (example from dairy cattle, Erbe et al., 2012). Using Goddard's (2009) theoretical approximation, $M_e = 2N_eL/(log(4N_eL))$, where L is the genome length, Meuwissen (2009) predicted that the training population size must approach $10 \times N_e \times L$ for an accuracy of 0.9 or $1 \times N_e \times L$ for an accuracy of 0.7 to 0.8 when training and test populations are unrelated.

Given the influence of N_e on the required training set size and number of markers, it is common to design prediction strategies that maximize the relationship between training and prediction populations and minimizes N_e to achieve high prediction accuracies with smaller training sets and fewer markers. The strength of LD observed between any marker—QTL pair will depend on the allele frequencies of the marker and QTL. To achieve maximum LD between the pairs, the frequencies of the major and minor alleles must be similar. Including individuals sharing the same alleles, for example half sibs, or merging of different groups of related populations guarantees increase in prediction accuracy for a given training set size and marker density. Biparental GS increases accuracy by maximizing the genetic relatedness of training and prediction sets, but requires phenotypes of lines from each cross before conducting GS, which may prolong the selection cycle and result in lower gains per year than multifamily GS proposed by Heffner et al. (2011). The multifamily approach is less 'population specific' and can leverage on the previous year's data for prediction accuracy. The downside to approaches that leverage population structure in outcrossing species is that the LD, and thus accuracy of prediction, deteriorates rapidly (Flint-Garcia et al., 2003), meaning the training populations must be continually "re-calibrated".

For any GS strategy it is critical to think about strategies for maintaining training sets that will yield optimal prediction accuracy given the resource constraints of a breeding program. For example, different breeding populations differ in the pattern of linkage between the marker and QTL. It is logical to think that with increasing marker density, the probability that some of the markers are close to the QTL and have consistent LD across breeds which can increase the prediction accuracy. It is, however, not about simply increasing the marker density (van den Berg et al., 2016; Raymond et al., 2018) but also about shared alleles, common ancestors (Goddard, 2017) and known QTL effects (van den Berg et al., 2016).

Some pre-cleaning of the data such as (i) removing SNPs with duplicate map positions, (ii) removing individuals with fewer than 90% of the SNP genotypes, (iii) removing individuals with excess heterozygosity, (iv) filtering SNPs with very low minor allele frequency, (v) filtering SNP with low imputation accuracy (Erbe et al., 2012), and (vi) removing one of the two adjacent markers that are in complete LD (Brøndum et al., 2012) can be adapted before analysis. Standardization of the marker effect also impacts the estimation as it induces less shrinkage to those effects with extreme (or intermediate) allele frequency (Gustavo de los Campos et al., 2013).

Intuition on GS model

Conceptually there are two useful ways to think about GS. The first is expressed by equation [1] in which the GEBV is a function of multiple regression on marker genotypes, and the second is to treat GEBV as a random effect encompassing all markers effects (GBLUP). When using a GBLUP approach the standard pedigree relationship matrix (**A**) is replaced with a genomic relationship matrix (**K**). The **K** matrix can be viewed as a realized relationship matrix whereas A is an expected relationship matrix. Further, Legarra et al. (2014) developed a single step model that combines pedigree and genomic information (**H**). The use of the **H** matrix enables analyses that utilize phenotypes from both genotypes and ungenotyped individuals. In this section I will provide an overview of several common approaches to calculate GEBV using both regression on marker genotypes and genomic relationship matrices.

Lets start by introducing a full model that includes both marker effects and a genomic relationship matrix (Kärkkäinen and Sillanpää, 2012):

$$\boldsymbol{y}_{n\times 1} = \mu + \boldsymbol{X}_{n\times p}\boldsymbol{\beta}_{p\times 1} + \boldsymbol{Z}_{n\times g}\boldsymbol{g}_{g\times 1} + \varepsilon_{n\times 1} \qquad [2]$$

where $\boldsymbol{y}$ is the vector of n observations; μ is the general mean; $\boldsymbol{X}$ and $\boldsymbol{\beta}$ are the incidence matrix and vector of marker effects respectively for p markers;

$\boldsymbol{Z}$ and $\boldsymbol{g}$ are the incidence matrix and vector of genotypic effects respectively of $\boldsymbol{g}$ genotypes; $\boldsymbol{\varepsilon}$ is the residual error. For a continuous observation, it is generally assumed that

$$\boldsymbol{g} \sim N(0, \boldsymbol{K}\sigma^2_{g*})$$
$$\varepsilon \sim N(0, \boldsymbol{I}\sigma^2)$$

where **K** is the genomic relationship matrix; * when a single variance is assumed across all the genotypes.

The full model is useful when there are a subset of markers tracking candidate genes that explain a relatively large portion of genetic variance. The effects of the candidate genes are modeled as fixed effects with the remaining polygenic variation captured by $\boldsymbol{g}$ and the **K** matrix. This modeling approach can also be used to identify candidate marker-QTL that are influencing the phenotype by calculating p-values for the markers in question while simultaneously controlling for background genetic effects. This model can also be further expanded to include effects for population structure (Lyra et al., 2018). Predictions from this model are calculated by summing the estimated marker effects with $\boldsymbol{g}$. It should be noted that markers must have a consistence linkage phase with the QTL in both the training and prediction sets, otherwise, the use of marker effect estimates will reduce prediction accuracy.

In a generalized form, the GS model using only the genomic relationship matrix can be written as,

$$y = \boldsymbol{X\beta} + \boldsymbol{Zu} + \varepsilon \qquad [3]$$

where $\boldsymbol{\beta}$ can be as simple as μ. The solution for the mixed effect model is as follows:

$$Var(y) = \hat{\boldsymbol{V}} = Z\hat{\boldsymbol{G}}\boldsymbol{Z}' + \hat{\boldsymbol{R}}$$
$$\hat{\boldsymbol{G}} = \boldsymbol{K}\sigma^2_g$$
$$\hat{\boldsymbol{R}} = \boldsymbol{I}\sigma^2$$

Prediction using the genotypic effect based on the relationship matrix is performed by matrix calculation as follows:

$$\hat{\boldsymbol{\beta}} = (\boldsymbol{X}'\,\hat{\boldsymbol{V}}^{-1}\,\boldsymbol{X})^{-1}\,\boldsymbol{X}'\,\hat{\boldsymbol{V}}^{-1}\boldsymbol{y}$$
$$\hat{\boldsymbol{u}} = \hat{\boldsymbol{G}}\boldsymbol{Z}'\,\hat{\boldsymbol{V}}^{-1}\,(y - \boldsymbol{X}\hat{\boldsymbol{\beta}})$$

The unknown parameter values are chosen based on the iterations to reach the global optimum. The logic of iteration is to start with some values for the parameters, keep changing the values at some gradient with a learning rate so that we can minimize the cost function (***J***). The ***J*** is some measure of goodness of fit which is used to minimize the difference between the estimated ($\hat{y}$) and observed (y) values. The gradient descent algorithm can be generalized as follows:

$$\begin{aligned} &\textit{repeat until convergence } \{ \\ &\hat{\boldsymbol{\beta}} = \hat{\boldsymbol{\beta}} - \alpha \boldsymbol{J}(y, \hat{y}) \\ &\hat{\boldsymbol{u}} = \hat{\boldsymbol{u}} - \alpha \boldsymbol{J}(y, \hat{y}) \\ &\} \end{aligned}$$

where convergence indicates minimization of $\boldsymbol{J}$; α is the learning rate. Both $\hat{\boldsymbol{\beta}}$ and $\hat{\boldsymbol{u}}$ must be updated simultaneously. Here, the unknown parameters to be updated are variance and mean. In other words, the parameter estimates are found by

$$(\hat{\boldsymbol{\beta}}, \hat{\boldsymbol{u}}) = l(y, \boldsymbol{\beta}, \boldsymbol{u})$$

where l denotes the maximum value of log likelihood. A model is fitted once the unknown parameters are chosen based on the highest likelihood. It is recommended that variance components be estimated using Restricted Maximum Likelihood (REML). The details of REML and algorithms for solving mixed model equations are considered out of scope for this chapter, but there are several packages available to solve mixed model equations.

Once we have the solution for the model using training data, it can be used for predictions in the test data. When the relationship matrix is used in the model, we need to multiply the solution with the incidence matrix of the test data while for models without a relationship matrix, coefficient values are added up to get the phenotype depending on the incidence matrix. The GS model with marker effects directly used is to find the QTL influencing the phenotype while that the genomic relationship matrix is used to rank genotypes and make selections for the breeding scheme.

Across-breed GS is a method developed in animal breeding to improve prediction accuracy in a mixture of breeding populations. The genetic similarity is estimated using a multi-breed genomic relationship matrix (MB-GRM) (Wientjes et al., 2017; 2018) in across-breed GS. On expanding the full model with the relationship matrix, a multivariate model with a single genomic relationship matrix, **K**, is calculated for multiple breeds. For example, for a two-breed population, the conventional GS model can be expanded as follows:

$$\begin{bmatrix} y_{breed1} \\ \hline y_{breed2} \end{bmatrix} = \begin{bmatrix} \mu_{breed1} \\ \hline \mu_{breed2} \end{bmatrix} + [\boldsymbol{Z}_{g1} \ \boldsymbol{Z}_{g2}] \begin{bmatrix} \boldsymbol{g}_1 \\ \hline \boldsymbol{g}_2 \end{bmatrix} + \begin{bmatrix} e_1 \\ \hline e_2 \end{bmatrix} \quad [4]$$

$$\begin{bmatrix} \boldsymbol{g}_1 \\ \hline \boldsymbol{g}_2 \end{bmatrix} \sim N(0, \boldsymbol{V} \otimes \boldsymbol{K}) \qquad \boldsymbol{V} = \left[\sigma^2_{g1} \ \sigma_{g1g2} \ \sigma_{g1g2} \ \sigma^2_{g2} \right]$$

where, $\mathbf{y}_{breed1}$ and $\mathbf{y}_{breed2}$ are the phenotypes for breed1 and breed2 respectively; μ_{breed1} and μ_{breed2} are the general means for breed1 and breed2 respectively;

$\mathbf{Z}_{g1}$ and $\mathbf{Z}_{g2}$ are the incidence matrices for breed1 and breed2 respectively; $\mathbf{g}_1$ and $\mathbf{g}_2$ are the genotypic effects for breed1 and breed2 respectively; e_1 and e_2 are the residual errors for breed1 and breed2 respectively; V is the variance covariance matrix for the genetic effect between two breeds. The model can be expanded to include multiple genomic relationship matrices. The genomic relationship matrix, $\boldsymbol{K}$, is calculated using all the marker information based on the method proposed by Wientjes et al. (2017) as follows:

$$K = \left[\frac{W_1 W'_1}{\sum 2q_{1_j}(1-q_{1_j})} \; \frac{W_1 W'_2}{\sqrt{\sum 2q_{1_j}(1-q_{1_j}) \sum 2q_{2_j}(1-q_{2_j})}} \; \frac{W_{g2} W'_{g1}}{\sqrt{\sum 2q_{1_j}(1-q_{1_j}) \sum 2q_{2_j}(1-q_{2_j})}} \; \frac{W_{g2} W'_{g2}}{\sum 2q_{2_j}(1-q_{2_j})} \right]$$

where, ***W1*** and ***W2*** are matrices for allele frequencies for all loci for breed1 and breed2 respectively; q is the allele frequency for locus j; the numbers 1 and 2 represent breed 1 and 2 respectively.

Genomic relationship matrix

The genomic relationship matrix used to estimate the GEBV is called a realized relationship matrix as it can capture the actual genomic segregation compared to the expected genetic relationship which is captured using a pedigree-based relationship matrix. The genomic relationship matrix can be calculated in multiple ways: additive or non-additive based on either the genetic theory or the Euclidean space among the genotypes using Gaussian or exponential functions (Endelman, 2011).

Calculation of the additive genomic relationship matrix (adopted from VanRaden, 2008) in a G-BLUP model is given below. Let M be a $n \times p$ matrix of marker information where n is the number of individuals and p is the number of markers. Let the frequency of the reference allele at locus j be q_j and let each column in $\mathbf{Q}$ contain the average allele count of the reference allele for each marker p expressed as $2q_j$.

$$\boldsymbol{W} = M - \boldsymbol{Q}$$

Subtraction of $\mathbf{Q}$ gives more credit to rare alleles than to common alleles when calculating genomic relationships. The genomic relationship matrix $\mathbf{K}$ which is scaled to be analogous to the pedigree relationship matrix, $\mathbf{A}$, such that the expectation of $\mathbf{K}$ is equal to the expectation of $\mathbf{A}$:

$$\boldsymbol{K} = \frac{\boldsymbol{WW}'}{2\sum q_j(1-q_j)}$$

The genomic inbreeding coefficient for individual i is simply $\boldsymbol{K}_{ii} - 1$, and the genomic relationships between individuals i and i' are obtained by dividing the elements $\boldsymbol{K}_{ii'}$ by square roots of diagonals $\boldsymbol{K}_{ii}$ and $\boldsymbol{K}_{i'i'}$.

A second method for calculating **K** (Leutenegger et al., 2003; Amin et al., 2007) is:

$$K = WDW'$$

where **D** is the diagonal with $D_{ii} = \frac{1}{p[2q_j(1-q_j)]}$

The third method to obtain **K** does not use allele frequencies but regresses **WW'** on **A** to obtain **K** using the model

$$WW' = g_0 + g_1 A + E$$

where g_0 and g_1 represent the intercept and slope; the matrix **E** includes differences of true from expected fractions of DNA because full DNA sequences were not available, and a subset of markers was genotyped instead. From the formula, **K** is obtained using

$$K = \frac{WW' - g_0}{g_1}$$

Frequentist methods

Penalized GS models – GBLUP, Ridge regression, LASSO, and Elastic net

The whole-genome regression (WGR) in a GS family of models directly uses a large number of markers (n << p issue) in the model. They can be generalized as follows:

$$y = \mu + X_{n \times p} \beta_{p \times 1} + \varepsilon_{n \times 1} \qquad [5]$$

where p is the number of markers. The WGR models are prone to overfitting and can also lead to difficulty in interpretation. The overfitting is due to the highly complex function defined specifically for the training dataset resulting in unreliable prediction accuracies on modifying the dataset. One method to control this issue is to use a regularization parameter, λ.

$$(\hat{\beta}, \hat{\mu}) = l(y, \beta, \mu) + \lambda(\beta)$$

The regularization parameter penalizes **β** keeping the function less sensitive to the training data. The regularization parameter (also called a tuning parameter or smoothing parameter) can be chosen by cross-validation (e.g., Craven and Wahba, 1978) or by Bayesian methods (e.g., Gianola and van Kaam, 2008). The iterative method adds a regularization parameter to the calculation of **β**.

$$\begin{aligned} &\textit{repeat until convergence } \{ \\ &\hat{\beta} = \hat{\beta} - \alpha[J(y, \hat{y}) + \lambda(\beta)] \\ &\hat{\mu} = \hat{\mu} - \alpha J(y, \hat{y}) \\ &\} \end{aligned}$$

Genomic best linear unbiased prediction (GBLUP) (VanRaden, 2008) is one of the earliest methods where a shrinkage parameter is used in GS. The GEBV obtained from the mixed model equation in BLUP (Henderson, 1975; Robinson, 1991) method is

$$GEBV = \boldsymbol{Z}(\boldsymbol{Z}'\,\boldsymbol{Z} + \boldsymbol{K}^{-1}\,\lambda)^{-1}\,\boldsymbol{Z}'(y - \boldsymbol{X}\hat{\boldsymbol{\beta}})$$

where $\lambda = \frac{\sigma_e^2}{\sigma_a^2}$ where σ_a^2 is the additive genetic variance. The λ in BLUP depends on the error and genetic variance which are assumed to be homogenous and so on the shrinkage parameter as well. In other words, the GEBV depends on the noise and estimated genetic variance.

Ridge regression (RR) (Hoerl and Kennard, 2000) which is equivalent to BLUP is another early model proposed for GS in a mixed model context using a regularization parameter. The solution by RR has improved numerical stability with highly correlated markers. With uncorrelated predictors, RR does proportional shrinkage, meaning the shrinkage is homogenous across markers. The ridge estimates cannot shrink the regression coefficients to zero. RR uses the *L2* form of the penalty, $\boldsymbol{J}(\boldsymbol{\beta}) = \sum_{h=1}^{p} ||\boldsymbol{\beta}_h||^2$. The general form of penalization comes from the formulation of bridge regression (Frank and Friedman, 1993) which uses $\boldsymbol{J}(\boldsymbol{\beta}) = \sum_{h=1}^{p} ||\boldsymbol{\beta}_h||^{\gamma}$. In RR, $\gamma = 2$, which cannot shrink the regression coefficients to zero.

Least Absolute Shrinkage and Selection Operator (LASSO) (Tibshirani, 1996) retains the goodness of both subset selection and RR. LASSO uses $\gamma = 1$ which is the *L1* form. Due to the differentially shrinking property of *L1* norm, LASSO can zero out some regression coefficients and can also provide shrinkage estimates for the remaining. A sparse model such as LASSO provides the potential to select a single or fewer markers that are highly correlated with the trait from the dense set of markers based on the assumption that QTL may be present in very few marker intervals. For example, LASSO proved to have higher accuracy compared to BLUP and BayesA in estimating the marker effect for traits which can be predicted using fewer markers in simulated mouse data (Graziano Usai et al., 2009).

There are a couple of limitations for *L1* and *L2* norms. First, in LASSO and other variable selection methods, the number of makers selected is restricted to *n* (Park and Casella, 2008; Mutshinda and Sillanpää, 2010). With complex traits such as yield, restricting the number of markers to *n* may lead to loss of information. Second, with a dense set of markers, it is possible that some of the markers are highly correlated to each other which can occur when linkage disequilibrium (LD) spans over long regions (Gustavo de los Campos et al., 2013). RR can deal with this multicollinearity while LASSO selects a single or fewer markers out of this group of correlated markers. On selecting a single marker out of a set of multicollinear markers may discard important information which leads to outperformance of RR over LASSO (Hastie et al., 2009). The later disadvantage is applicable in scenarios where number of QTL is more than the effective population size (Heffner et al., 2011).

To retain the goodness of LASSO and RR, Zou and Hastie (2005) proposed Elastic Net (EN) as a penalty with a weighted average of the *L1* and *L2* norms. The EN has two tuning parameters – λ and ζ – in the norm $0 \leq \zeta \leq 1$, $\boldsymbol{J(\beta)} = \zeta \sum_{h=1}^{p} ||\boldsymbol{\beta}_h|| + (1 - \zeta)\sum_{h=1}^{p} ||\boldsymbol{\beta}_h||^2$. EN has the desirable property of LASSO that it can shrink the regression coefficients to zero. It has an advantage over LASSO that it tends to select all the multicollinear markers to the model and assigns equal regression coefficients to them.

Kernel based methods – Reproducing Kernel Hilbert Space (RKHS)

RKHS is a regularization class of statistical models where a kernel is used for optimization. RKHS is considered as a special case scenario of BLUP in animal breeding (Morota and Gianola, 2014). In RKHS, the genomic and/or pedigree-based relationship matrix can be used in the model. The pedigree-based matrix is formed on the theory of identical by descent (ibd) and is modeled using a parametric kernel. On the other hand, the genomic matrix based on the genomic state of the individuals (identical by state (ibs)) is less dependent on the ancestry but more on the shared alleles. Thus, the ibs based kernels can be modeled using parametric as well as non-parametric methods such as distance-based approaches allowing more flexibility in evaluation (Morota and Gianola, 2014; Shawe-Taylor and Cristianini, 2004; Endelman, 2011). RKHS was found to perform in prediction accuracy on comparing 10 different methods for 18 different traits in barley, wheat and maize (Heslot et al., 2012). RKHS can take in not just the additive effect but also non-additive effects such as the dominance effect (Wellmann and Bennewitz, 2012).

For convenience, the WGR model which can include a genomic relationship (and a pedigree relationship matrix) can be re-written as (adopted from de los Campos et al., 2009)

$$y = \boldsymbol{f} + \varepsilon \qquad [6]$$

where, $\boldsymbol{f}$ can be viewed as a vector, $\boldsymbol{f} = (f_1, ..., f_p)'$, where each entry gives the evaluation of the function $\boldsymbol{f}$ at a point in the set of predictors, P. In RKHS, a semi parametric method, the penalty can now be considered as $\boldsymbol{J(f)}$. Therefore, the optimization in RKHS can be expressed as

$$\hat{\boldsymbol{f}}(P) = \{l(y, \boldsymbol{f}, P) + \lambda||\boldsymbol{f}||_H^2\}$$

In RKHS, it is assumed that $\boldsymbol{f}$ belongs to a Hilbert space, H (Akhiezer and Glazman, 1993), and the square of the norm $\boldsymbol{f}$ in that space is used for regularization; $||f||_H$ denotes the norm in H. Model specifications in RKHS regression refers to the choice of the cost function, ***J***(), *H*, and λ. The choice of *H* is central in RKHS defining the space over which ***f*** is performed.

Hilbert space (Hilbert et al., 1928), *H*, is a linear vector space which satisfies the conditions of the inner product operations of two vectors. The inner product is a way to multiply vectors together with the result being a scalar. A Hilbert space is separable meaning it contains countable dense objects. Rational numbers (the numbers that can be written in the form of a ratio also) are examples of countable objects. Dense means any rational number that is arbitrarily close to the irrational number can be found. Closeness is a condition of the inner product operation of two vectors that the distance between any two vectors is defined as the square root of the inner product of the difference in the vectors. As the Hilbert space is dense, it contains set of real numbers ($\mathbb{R}$) (both rational and irrational) without any space between them. In other words, there is no gap in the Hilbert space or it is complete.

A Hilbert space is a RKHS when the functional kernels in the space are bounded and continuous (no gap). The kernels are bounded meaning an upper and lower boundary values are given. Instead of vectors, kernels are used in RKHS. These kernels also satisfy the positive definite condition as that of vectors. On knowing the kernel and its boundary limits the RKHS can be reproduced meaning it can be evaluated on a continuous domain in the Hilbert space (Aronszajn, 1950).

For every RKHS, there exists an associated reproducing kernel (G) which is symmetric and positive definite or semidefinite. Selecting a kernel is the most critical stage in applying RKHS (Shawe-Taylor and Cristianini, 2004). The choice of kernel is not just based on prior information (as in Bayesian models) but also an estimation problem (as in choosing the best kernel and its parameters through a grid search). A commonly used symmetric positive definite kernel in GS is Gaussian kernel.

$$G(x,x') = e^{-\frac{\|x-x'\|^2}{\sigma^2}}, \quad \sigma > 0$$

where $x \in X$, X be some set, for example $\mathbb{R}$ or a subset of $\mathbb{R}$.

Bayesian methods

Bayesian methods provide a framework for including prior information on marker effects. This can be beneficial when dealing with sequence level information, by applying prior knowledge on gene space (i.e., which markers are in coding regions of the genome). Various types of Bayesian priors assume marker effects are samples from multiple distributions, resulting in heavily shrunk estimates with little or no effect but those lightly shrunk with larger effects. Bayesian models work better when major QTL are in strong LD with markers (Desta and Ortiz, 2014). Association studies and comparative studies with other GS models help in understanding the strength of LD. Consequently, Bayesian models are preferred when number of QTL < M_e. On the other hand, when QTL > M_e, or effectively when all the QTL have small effects, differential shrinkage does not hold and models that apply uniform shrinkage tend to perform well relative to mixture models. Bayesian methods with proper priors induce regularization or penalty automatically which tackles the n << p problem making the number of non-zero coefficients < n. So, as n grows the influence of regularization diminishes. However, the inferences on the genetic architecture always depend on the priors adopted; in other words, the posterior distribution is dependent on the a priori distribution (Gianola, 2013; Kärkkäinen and Sillanpää, 2012).

Unlike frequentist methods, the Bayesian approach is not aiming to determine the estimates alone, but its distribution—posterior distribution. Thus, improvement in prediction accuracy depends on the probability distribution which varies based on the priors. The priors should be specified based on the assumed genetic architecture of the trait and known sources of variation in the data (Kärkkäinen and Sillanpää, 2012). The general structure of Bayesian linear models used in GS is (adopted from Gustavo de los Campos et al., 2013),

$$p(\mu, \boldsymbol{\beta}, \sigma^2 | \boldsymbol{y}, \boldsymbol{\omega}) \qquad [7]$$
$$\propto p(\boldsymbol{y}|\mu, \boldsymbol{\beta}, \sigma^2)\, p(\mu, \boldsymbol{\beta}, \sigma^2|\boldsymbol{\omega})$$
$$\propto \prod_{i=1}^{n} N\left(y_i \mid \mu + \sum_{h=1}^{p} x_{ih}\beta_h, \sigma^2\right) \prod_{h=1}^{p} p(\beta_h \mid \boldsymbol{\omega}) p(\sigma^2)$$

where $p(\mu, \boldsymbol{\beta}, \sigma^2|\boldsymbol{y}, \boldsymbol{\omega})$ is the posterior density of unknown parameters $\{\mu, \boldsymbol{\beta}, \sigma^2\}$ given the data $\boldsymbol{y}$ and hyperparameters $\boldsymbol{\omega}$. Here, $\boldsymbol{\beta}$ is the effect of predictors, and σ^2 is the residual variance. The posterior density represents our prior belief about the value of the parameter, mostly based on assumptions, or historical information. Hyperparameters are for tuning the model. Examples are learning rate α, optimization parameter λ, degrees of freedom and variance. The posterior density defines the type of model in the Bayesian family of models. Based on the distribution of mass around zero and tails, de los Campos et al. (2013) classified the most used priors into four categories: Gaussian, thick tail, spike – slab, and point of mass—which are briefed in the following paragraphs.

Gaussian prior

The Gaussian prior has two hyperparameters: the mean (commonly set to zero) and the variance σ^2_β. With a known general mean (intercept) and variance parameters, the posterior distribution of marker effects is,

$$p(\boldsymbol{\beta}|\boldsymbol{y}, \mu, \sigma^2, \sigma^2_\beta) \qquad [8]$$

$$\propto \prod_{i=1}^{n} N\left(y_i \mid \mu + \sum_{h=1}^{p} x_{ih}\beta_h, \sigma^2\right) \prod_{h=1}^{p} N(\beta_h \mid 0, \sigma_\beta^2)$$

The posterior distribution is multivariate normal with a mean similar to that of BLUP and RR estimates. Therefore, the model is referred to as Bayesian ridge regression (BRR) or RR-BLUP. Like RR, BRR performs homogenous shrinkage across the markers. Therefore, BRR is not a preferred choice when differential shrinkage is required.

Thick tailed prior

The two most commonly used densities in GS under this category are scaled *t* as in Bayes A (Meuwissen et al., 2001) and double exponential (DE) as in Bayesian LASSO (BL) (Park and Casella, 2008). Just like in LASSO, BL also offers differential shrinkage: strong shrinkage towards zero for estimates with smaller effects while less shrinkage for large effects. The characteristics of BL are similar to those of LASSO in dealing with sparse models with a striking similarity in the results (Park and Casella, 2008). The prior density is represented as infinite mixtures of scaled normal densities (Andrews and Mallows, 1974). When the prior density $p(\sigma_\beta^2|\boldsymbol{\omega})$ assigned to the markers is a scaled inverse chi-square density it is called scaled *t* and when it is an exponential density it is called DE. The DE has one parameter while student *t* has two parameters to tune.

Mutshinda and Silanappa (2010) proposed an extension of BL (EBL). Compared to having a single regularization hyperparameter as in BL, EBL has two regularization parameters: delta, a measure of model sparsity and eta, a locus specific deviation from delta. With the two regularization parameters, the differential shrinkage is more prominent in EBL and it is found to be outperforming BL in prediction accuracy.

The spike-slab prior

The prior density uses a mixture of normal densities: one with small variance (the spike) and the other with large variance (the slab) (e.g., George and McCulloch, 1993). The prior density can be written as

$$p(\boldsymbol{\beta}|\pi, \sigma_{\beta 1}^2, \sigma_{\beta 2}^2) = \pi \times N(\boldsymbol{\beta}|0, \sigma_{\beta 1}^2) + (1 - \pi) \times N(\boldsymbol{\beta}|0, \sigma_{\beta 2}^2) \quad [9]$$

where $\pi \in [0,1]$ and $\sigma_{\beta 1}^2$ and $\sigma_{\beta 2}^2$ are variance parameters where $\sigma_{\beta 1}^2$ represents the spike, $\sigma_{\beta 1}^2 \leq \sigma_{\beta 2}^2$. In the Bayesian GS model stochastic search variable selection (SSVS) (Calus et al., 2008; Verbyla et al., 2009), a reparameterization of the prior is used so that the variance of one of the components is a scaled version of the variance of the other component. SSVS assumes that markers with negligible effects come from the distribution with a small variance (Wellmann and Bennewitz, 2012).

Point of mass at zero and slab priors

These kinds of priors are used to induce a combination of variable selection and shrinkage. These are used for example in models Bayes B with scaled *t* slab (Meuwissen et al., 2001) which can be considered as a limiting case of SSVC and Bayes C with a normal density slab (David Habier et al., 2011). Meuwissen (2009) noticed that superiority of BayesB over GBLUP increased with increasing marker density.

A decrease in the number of QTL per sample size is proportional to the outperformance of BayesB and in the reverse scenario linear regression outperforms. On comparing the variable selection approaches in WGR such as LASSO, elastic net, and Bayes B vs RR-BLUP it is found that under scenarios of low LD, high heritability, large sample size, and lower casual mutation relative to sample size, variable selection methods had better accuracy than RR-BLUP; otherwise the reverse is true (Desta and Ortiz, 2014).

Other deviations of Bayesian

With reasonable modification in priors, it can be shown that models such as BRR, BayesA, and Bayes C are special cases of Bayes B. Priors for Bayes A and B follows the assumption that some markers have no effect on the trait of interest with a probability π and a subset of markers contributing to the trait have some effect sampled from a scaled t distribution with probability $(1 - \pi)$. The models assume that the probability is known in both scenarios: in Bayes A, $\pi = 0$ assuming there are no markers with zero effect while in Bayes B the assuming that there are many markers with zero effect treating $\pi > 0$. However, the given shrinkage depends on this hyperparameter, it should be treated as unknown and inferred from the data. Bayes C π and D π (David Habier et al., 2011) are versions of Bayes C and D respectively when π is treated as unknown. Bayes C considers a single effect variance for all the markers unlike the locus specific variance in Bayes A and B. In Bayes D the regularization parameter is treated as unknown. Apart from these concerns and its modifications, Bayes A and B are still the popular choices because their implementation as locus samplers is straightforward and computing time is reasonable (Meuwissen et al., 2001; Habier et al., 2007; 2009).

Tuning hyperparameters

Choice of hyperparameters is data dependent. There are many methods to choose the values for hyperparameters: (i) based on a prior expectation of the genetic variance (Gustavo de los Campos et al., 2013), (ii) fit the model over a grid of values of $\boldsymbol{\omega}$ by randomly choosing values from the grid. A random choice of values from the grid is recommended because it is computationally less demanding, and some of the parameters are more important than others.

For example, in a systematic examination of the grid, we might be repeating the results of an important parameter by keeping its value the same while changing the value of other parameters. This repetition can be minimized by randomly choosing the values from the grid.

A great advantage of Bayesian models is that we can potentially customize and include any prior information into the model. For example, we can include prior information (i) if the markers are in independent and identically distributed (iid) paving the way for homogenous shrinkage, (ii) location of the marker—whether the marker is in the coding or non-coding region, (iii) whether the marker is tightly linked to the QTL, (iv) different priors for different sets of markers (Gustavo de los Campos et al., 2013; Gianola, 2013). Based on these hyperparameters, modifications in shrinkage and estimates are done and inferences are made. There are algorithms to approximate the posterior distribution based on this prior information.

Algorithms

A commonly used algorithm in Bayesian methods is Monte Carlo Marko Chain (MCMC) (Gelman et al., 2014). MCMC is an algorithm to sample from a probability distribution. MCMC is used to build Bayesian models because Bayesian statistics are built on the assumption that probability of an incident depends on its prior assumption and the likelihood of that incident based on the information from the data (van de Schoot et al., 2021). In MCMC, a chain of steps is constructed for the desired probability distribution. Each step is based on the probability of the previous incident. The more the steps the more accurate the sampling of the desired distribution. An example of MCMC is Gibbs samples. A description of the steps in the Gibbs sampler is given by de los Campos et al. (2013).

However, MCMC can be computationally demanding with large datasets as the cost scales up at $\omega(n)$ per step (Cornish et al., 2019). There are other algorithms such as Expectation maximization (EM) (Dempster et al., 1977) which have proved to be faster and can be used in larger datasets. The EM algorithm is an iterative method to estimate ML where values for the parameters are estimated first, then optimization of the model is performed. The iteration continues until convergence. The fast EM algorithms are proposed for BL (Casella, 2001), Bayes A called, fastBayes A (Sun et al., 2012), and for Bayes B, fastBayes B (Meuwissen et al., 2009).

Some more extended Bayesian models

There are Bayesian models considering non-additive effects in addition to the additive effects. Bayes D is such a model considering dominance effect (Wellmann and Bennewitz, 2012). Bayes D shows superiority to models with additive effects only when markers having strong LD with QTL that can capture dominance are also present. The strong LD maintains the prediction accuracy which can otherwise degenerate due to recombination. Over generations, however, the prediction accuracy on including the dominance effect decreases.

There are other Bayesian models which partition the chromosomes into smaller segments (e.g., 100 markers each) and provide different prior probabilities to different segments. This type of modeling assumes that some part of the genome explains more variation than others. Bayes R (Erbe et al., 2012) and Bayes RS (Brøndum et al., 2012) are examples of such models. Assuming that the same causative mutations, or even the same gene regions but different causative mutations, act on traits of interest in different populations, it is expected that effects of chromosome regions on a trait could be consistent among populations, though, the LD patterns between individual SNPs and QTLs could differ from one population to the other (Brøndum et al., 2012). Further investigations, however, are required to confirm the robustness and usefulness of such segment-wise models.

Conclusions

Rather than following a trait controlled by a single or a very few loci as in MAS, GS opens a broad road in the breeding scheme design with the potential to predict quantitative traits involving multiple large and small effect loci. GS based selection also offers the flexibility of testing a large number of genotypes where alleles of interest are replicated and not the genotype itself. The prediction accuracy in GS models, however, depends on several factors. Some of these factors such as heritability, genetic architecture, and to a large extent LD cannot be controlled; however, we can control the design of data sets including size and relationships, marker density, and the type of model (de los Campos et al., 2013). With GBS becoming more and more affordable and with new advances in modeling and computation, GS will be a critically important tool in plant breeding pipelines for complex traits.

Acknowledgments

I am thankful to the Department of Biotechnology (DBT), Government of India, for granting me the Ramalingaswami fellowship to continue my research work in GS which helped me to write this chapter. I am also thankful to Dr. Kelly R. Robbins, Cornell University, USA, for his critical review of this chapter.

References

Akhiezer, N. I. and Glazman, I. M. 1993. *Theory of Linear Operators in Hilbert Space*. New York: Dover Publications.

Amin, Najaf, Cornelia M. van Duijn and Yurii S. Aulchenko. 2007. A genomic background based method for association analysis in related individuals. Edited by Peter Heutink. *PLoS ONE* 2(12): e1274. https://doi.org/10.1371/journal.pone.0001274.

Andrews, D. F. and Mallows, C. L. 1974. Scale mixtures of normal distributions. *Journal of the Royal Statistical Society: Series B (Methodological)* 36(1): 99–102. https://doi.org/10.1111/j.2517-6161.1974.tb00989.x.

Aronszajn, N. 1950. Theory of reproducing kernels. *Transactions of the American Mathematical Society* 68(3): 337–337. https://doi.org/10.1090/S0002-9947-1950-0051437-7.

Arruda, M. P., Lipka, A. E., Brown, P. J., Krill, A. M., Thurber, C. et al. 2016. Comparing genomic selection and marker-assisted selection for fusarium head blight resistance in wheat (*Triticum aestivum* L.). *Molecular Breeding* 36(7): 84. https://doi.org/10.1007/s11032-016-0508-5.

Berg, Irene van den, Didier Boichard, Bernt Guldbrandtsen, Mogens S. Lund et al. 2016. Using sequence variants in linkage disequilibrium with causative mutations to improve across-breed prediction in dairy cattle: a simulation study. *G3 Genes|Genomes|Genetics* 6(8): 2553–61. https://doi.org/10.1534/g3.116.027730.

Bernardo, Rex. 2008. Molecular markers and selection for complex traits in plants: learning from the last 20 years. *Crop Science* 48(5): 1649–64. https://doi.org/10.2135/cropsci2008.03.0131.

Brøndum, Rasmus, Guosheng Su, Mogens Lund, Philip J. Bowman et al. 2012. Genome position specific priors for genomic prediction. *BMC Genomics* 13(1): 543. https://doi.org/10.1186/1471-2164-13-543.

Calus, M. P. L., Meuwissen, T. H. E., de Roos, A. P. W. and Veerkamp, R. F. 2008. Accuracy of genomic selection using different methods to define haplotypes. *Genetics* 178(1): 553–61. https://doi.org/10.1534/genetics.107.080838.

Campos, G. de los, Gianola, D. and Rosa, G. J. M. 2009. Reproducing kernel hilbert spaces regression: a general framework for genetic evaluation1. *Journal of Animal Science* 87(6): 1883–87. https://doi.org/10.2527/jas.2008-1259.

Campos, Gustavo de los, John M. Hickey, Ricardo Pong-Wong, Hans D. Daetwyler and Mario P. L. Calus et al. 2013. Whole-genome regression and prediction methods applied to plant and animal breeding. *Genetics* 193(2): 327–45. https://doi.org/10.1534/genetics.112.143313.

Casella, G. 2001. Empirical Bayes Gibbs sampling. *Biostatistics* 2(4): 485–500. https://doi.org/10.1093/biostatistics/2.4.485.

Cornish, Rob, Paul Vanetti, Alexandre Bouchard-Cote, George Deligiannidis and Arnaud Doucet et al. 2019. Scalable metropolis-hastings for exact Bayesian inference with large datasets. *In*: *Proceedings of the 36th International Conference on Machine Learning*. Long Beach, California.

Craven, Peter and Grace Wahba. 1978. Smoothing noisy data with spline functions: estimating the correct degree of smoothing by the method of generalized cross-validation. *Numerische Mathematik* 31(4): 377–403. https://doi.org/10.1007/BF01404567.

Daetwyler, Hans D., Ricardo Pong-Wong, Beatriz Villanueva and John A. Woolliams. 2010. The impact of genetic architecture on genome-wide evaluation methods. *Genetics* 185(3): 1021–31. https://doi.org/10.1534/genetics.110.116855.

Dempster, A. P., Laird, N. M. and Rubin, D. B. 1977. Maximum likelihood from incomplete data via the *EM* algorithm. *Journal of the Royal Statistical Society: Series B (Methodological)* 39(1): 1–22. https://doi.org/10.1111/j.2517-6161.1977.tb01600.x.

Desta, Zeratsion Abera and Rodomiro Ortiz. 2014. Genomic selection: genome-wide prediction in plant improvement. *Trends in Plant Science* 19(9): 592–601. https://doi.org/10.1016/j.tplants.2014.05.006.

Endelman, Jeffrey B. 2011. Ridge regression and other kernels for genomic selection with R Package RrBLUP. *The Plant Genome* 4(3): 250–55. https://doi.org/10.3835/plantgenome2011.08.0024.

Erbe, M., Hayes, B. J., Matukumalli, L. K., Goswami, S., Bowman, P. J. et al. 2012. Improving accuracy of genomic predictions within and between dairy cattle breeds with imputed high-density single nucleotide polymorphism panels. *Journal of Dairy Science* 95(7): 4114–29. https://doi.org/10.3168/jds.2011-5019.

Falconer, Douglas S. and Trudy Mackay. 2009. *Introduction to Quantitative Genetics*. 4. ed., [16. print.]. Harlow: Pearson, Prentice Hall.

Flint-Garcia, Sherry A., Jeffry M. Thornsberry and Edward S. Buckler. 2003. Structure of linkage disequilibrium in plants. *Annual Review of Plant Biology* 54(1): 357–74. https://doi.org/10.1146/annurev.arplant.54.031902.134907.

Frank, lldiko E. and Jerome H. Friedman. 1993. A statistical view of some chemometrics regression tools. *Technometrics* 35(2): 109–35. https://doi.org/10.1080/00401706.1993.10485033.

Gelman, Andrew, John B. Carlin, Hal Steven Stern, David B. Dunson, Aki Vehtari et al. 2014. *Bayesian Data Analysis*.

George, Edward I. and Robert E. McCulloch. 1993. Variable selection via Gibbs sampling. *Journal of the American Statistical Association* 88(423): 881–89. https://doi.org/10.1080/01621459.1993.10476353.

Gianola, Daniel and Johannes B. C. H. M. van Kaam. 2008. Reproducing kernel hilbert spaces regression methods for genomic assisted prediction of quantitative traits. *Genetics* 178(4): 2289–2303. https://doi.org/10.1534/genetics.107.084285.

Gianola, Daniel. 2013. Priors in whole-genome regression: the Bayesian alphabet returns. *Genetics* 194(3): 573–96. https://doi.org/10.1534/genetics.113.151753.

Goddard, M. E. 2017. Can we make genomic selection 100% accurate? *Journal of Animal Breeding and Genetics* 134(4): 287–88. https://doi.org/10.1111/jbg.12281.

Goddard, Mike. 2009. Genomic selection: prediction of accuracy and maximisation of long term response. *Genetica* 136(2): 245–57. https://doi.org/10.1007/s10709-008-9308-0.

Graziano Usai, Mario, Mike Goddard and Ben Hayes. 2009. Using LASSO to estimate marker effects for genomic selection. *Italian Journal of Animal Science* 8(sup2): 168–70. https://doi.org/10.4081/ijas.2009.s2.168.

Habier, D., Fernando, R. L. and Dekkers, J. C. M. 2007. The impact of genetic relationship information on genome-assisted breeding values. *Genetics* 177(4): 2389–97. https://doi.org/10.1534/genetics.107.081190.

Habier, D., Fernando, R. L. and Dekkers, J. C. M. 2009. Genomic selection using low-density marker panels. *Genetics* 182(1): 343–53. https://doi.org/10.1534/genetics.108.100289.

Habier, David, Rohan L. Fernando, Kadir Kizilkaya and Dorian J. Garrick. 2011. Extension of the Bayesian alphabet for genomic selection. *BMC Bioinformatics* 12(1): 186. https://doi.org/10.1186/1471-2105-12-186.

Hastie, Trevor, Robert Tibshirani and Friedman, J. H. 2009. *The Elements of Statistical Learning: Data Mining, Inference, and Prediction*. 2nd ed. Springer Series in Statistics. New York, NY: Springer.

Hayes, B. J., Visscher, P. M. and Goddard, M. E. 2009. Increased accuracy of artificial selection by using the realized relationship matrix. *Genetics Research* 91(1): 47–60. https://doi.org/10.1017/S0016672308009981.

Heffner, Elliot L., Jean-Luc Jannink and Mark E. Sorrells. 2011. Genomic selection accuracy using multifamily prediction models in a wheat breeding program. *The Plant Genome* 4(1): 65–75. https://doi.org/10.3835/plantgenome2010.12.0029.

Henderson, C. R. 1975. Best linear unbiased estimation and prediction under a selection model. *Biometrics* 31(2): 423–47.

Heslot, Nicolas, Hsiao-Pei Yang, Mark E. Sorrells and Jean-Luc Jannink. 2012. Genomic selection in plant breeding: a comparison of models. *Crop Science* 52(1): 146–60. https://doi.org/10.2135/cropsci2011.06.0297.

Hilbert, D., Neumann, J. v. and Nordheim, L. 1928. Uber die Grundlagen der Quantenmechanik. *Mathematische Annalen* 98(1): 1–30. https://doi.org/10.1007/BF01451579.

Hill, G. William. 2012. Quantitative genetics in the genomics era. *Current Genomics* 13(3): 196–206. https://doi.org/10.2174/138920212800543110.

Hoerl, Arthur E. and Robert W. Kennard. 2000. Ridge regression: biased estimation for nonorthogonal problems. *Technometrics* 42(1): 80–86. https://doi.org/10.1080/00401706.2000.10485983.

Hoffstetter, Amber, Antonio Cabrera, Mao Huang and Clay Sneller. 2016. Optimizing training population data and validation of genomic selection for economic traits in soft winter wheat. *G3 (Bethesda, Md.)* 6(9): 2919–28. https://doi.org/10.1534/g3.116.032532.

Jannink, J.-L., Lorenz, A. J. and Iwata, H. 2010. Genomic selection in plant breeding: from theory to practice. *Briefings in Functional Genomics* 9(2): 166–77. https://doi.org/10.1093/bfgp/elq001.

Kärkkäinen, Hanni P. and Mikko J. Sillanpää. 2012. Back to basics for bayesian model building in genomic selection. *Genetics* 191(3): 969–87. https://doi.org/10.1534/genetics.112.139014.

Kemper, K. E., Littlejohn, M. D., Lopdell, T., Hayes, B. J., Bennett, L. E. et al. 2016. Leveraging genetically simple traits to identify small-effect variants for complex phenotypes. *BMC Genomics* 17(1): 858. https://doi.org/10.1186/s12864-016-3175-3.

Knapp, S. J. and Bridges, W. C. 1990. Using molecular markers to estimate quantitative trait locus parameters: power and genetic variances for unreplicated and replicated progeny. *Genetics* 126(3): 769–77.

Legarra, Andres, Ole F. Christensen, Ignacio Aguilar and Ignacy Misztal. 2014. Single step, a general approach for genomic selection. *Livestock Science* 166(August): 54–65. https://doi.org/10.1016/j.livsci.2014.04.029.

Leutenegger, Anne-Louise, Bernard Prum, Emmanuelle Génin, Christophe Verny, Arnaud Lemainque et al. 2003. Estimation of the inbreeding coefficient through use of genomic data. *American Journal of Human Genetics* 73(3): 516–23. https://doi.org/10.1086/378207.

Lyra, Danilo Hottis, Ítalo Stefanine Correia Granato, Pedro Patric Pinho Morais, Filipe Couto Alves et al. 2018. Controlling population structure in the genomic prediction of tropical maize hybrids. *Molecular Breeding* 38(10): 126. https://doi.org/10.1007/s11032-018-0882-2.

Meuwissen, T. H., Hayes, B. J. and Goddard, M. E. 2001. Prediction of total genetic value using genome-wide dense marker maps. *Genetics* 157(4): 1819–29.

Meuwissen, Theo H. E. 2009. Accuracy of breeding values of ‘unrelated’ individuals predicted by dense SNP genotyping. *Genetics Selection Evolution* 41(1): 35. https://doi.org/10.1186/1297-9686-41-35.

Meuwissen, Theo H. E., Trygve R. Solberg, Ross Shepherd and John A. Woolliams et al. 2009. A fast algorithm for BayesB type of prediction of genome-wide estimates of genetic value. *Genetics Selection Evolution* 41(1): 2. https://doi.org/10.1186/1297-9686-41-2.

Morota, Gota and Daniel Gianola. 2014. Kernel-based whole-genome prediction of complex traits: a review. *Frontiers in Genetics* 5(October). https://doi.org/10.3389/fgene.2014.00363.

Mutshinda, Crispin M. and Mikko J. Sillanpää. 2010. Extended Bayesian LASSO for multiple quantitative trait loci mapping and unobserved phenotype prediction. *Genetics* 186(3): 1067–75. https://doi.org/10.1534/genetics.110.119586.

Park, Trevor and George Casella. 2008. The Bayesian Lasso. *Journal of the American Statistical Association* 103(482): 681–86. https://doi.org/10.1198/016214508000000337.

Raymond, Biaty, Aniek C. Bouwman, Chris Schrooten, Jeanine Houwing-Duistermaat, Roel F. Veerkamp et al. 2018. Utility of whole-genome sequence data for across-breed genomic prediction. *Genetics Selection Evolution* 50(1): 27. https://doi.org/10.1186/s12711-018-0396-8.

Robinson, G. K. 1991. That BLUP is a good thing: the estimation of random effects. *Statistical Science* 6(1): 15–32. https://doi.org/10.1214/ss/1177011926.

Schoot, Rens van de, Sarah Depaoli, Ruth King, Bianca Kramer, Kaspar Märtens et al. 2021. Bayesian statistics and modelling. *Nature Reviews Methods Primers* 1(1): 1. https://doi.org/10.1038/s43586-020-00001-2.

Shamshad, Mohd and Achla Sharma. 2018. The usage of genomic selection strategy in plant breeding. *In*: *Next Generation Plant Breeding*, edited by Yelda Özden Çiftçi. InTech. https://doi.org/10.5772/intechopen.76247.

Shawe-Taylor, John and Nello Cristianini. 2004. *Kernel Methods for Pattern Analysis*. Cambridge, UK; New York: Cambridge University Press.

Solberg, T. R., Sonesson, A. K., Woolliams, J. A. and Meuwissen, T. H. E. 2008. Genomic selection using different marker types and densities. *Journal of Animal Science* 86(10): 2447–54. https://doi.org/10.2527/jas.2007-0010.

Sun, Xiaochen, Long Qu, Dorian J. Garrick, Jack C. M. Dekkers, Rohan L. Fernando et al. 2012. A fast EM algorithm for BayesA-Like prediction of genomic breeding values. Edited by Rongling Wu. *PLoS ONE* 7(11): e49157. https://doi.org/10.1371/journal.pone.0049157.

Tibshirani, Robert. 1996. Regression shrinkage and selection via the Lasso. *Journal of the Royal Statistical Society. Series B (Methodological)* 58(1): 267–88.

VanRaden, P.M. 2008. Efficient methods to compute genomic predictions. *Journal of Dairy Science* 91(11): 4414–23. https://doi.org/10.3168/jds.2007-0980.

Verbyla, Klara L., Ben J. Hayes, Philip J. Bowman, Michael E. Goddard et al. 2009. Accuracy of genomic selection using stochastic search variable selection in australian holstein friesian dairy cattle. *Genetics Research* 91(5): 307–11. https://doi.org/10.1017/S0016672309990243.

Villar-Hernández, Bartolo de Jesús, Sergio Pérez-Elizalde, José Crossa, Paulino Pérez-Rodríguez et al. 2018. A Bayesian decision theory approach for genomic selection. *G3: Genes|Genomes|Genetics* 8(9): 3019–37. https://doi.org/10.1534/g3.118.200430.

Wellmann, Robin and Jörn Bennewitz. 2012. Bayesian models with dominance effects for genomic evaluation of quantitative traits. *Genetics Research* 94(1): 21–37. https://doi.org/10.1017/S0016672312000018.

Wientjes, Yvonne C. J., Piter Bijma, Jérémie Vandenplas, Mario P. L. Calus et al. 2017. Multi-population genomic relationships for estimating current genetic variances within and genetic correlations between populations. *Genetics* 207(2): 503–15. https://doi.org/10.1534/genetics.117.300152.

Wientjes, Yvonne C. J., Mario P. L. Calus, Pascal Duenk et al. 2018. Required properties for markers used to calculate unbiased estimates of the genetic correlation between populations. *Genetics Selection Evolution* 50(1): 65. https://doi.org/10.1186/s12711-018-0434-6.

Wray, Naomi R., Michael E. Goddard and Peter M. Visscher. 2007. Prediction of individual genetic risk to disease from genome-wide association studies. *Genome Research* 17(10): 1520–28. https://doi.org/10.1101/gr.6665407.

Zou, Hui and Trevor Hastie. 2005. Regularization and variable selection via the elastic net. *Journal of the Royal Statistical Society. Series B (Statistical Methodology)* 67(2): 301–20.

2

Hands on Training Optimization in Genomic Selection

Isidro y Sánchez Julio,[1] *Rio Simon*[1] and *Akdemir Deniz*[2]

ABSTRACT

Genomic selection (GS) is a tool in plant and animal breeding that utilizes machine learning approaches to make predictions of un-phenotyped individuals to make selection decisions. Constructing genomic prediction models requires genome-wide marker data along with phenotypic data to build a training population set (TRS). The selection of the TRS is critical for the success of GS since the predictions are based on markers or individual effects estimated on the TRS. Here, we review the different criteria proposed in the literature when designing a TRS. In addition, we provide a practical overview of the statistical analysis needed to optimize the TRS using R. The statistical procedure is performed by the R package TrainSel, and issues associated with the analysis are addressed along with the R code. The ultimate aim of this chapter is to provide a practical guideline to perform TRS optimization analysis using R, rather than describe the theory in depth.

Brief introduction on genomic selection

The big picture in plant breeding focuses on three main steps: crossing, evaluation, and selection. Crossing elite material evaluating the progenies

[1] Centro de Biotecnología y Genómica de Plantas (CBGP, UPM-INIA) Universidad Politécnica de Madrid (UPM) - Instituto Nacional de Investigación y Tecnología Agraria y Alimentaria (INIA) Campus de Montegancedo-UPM 28223-Pozuelo de Alarcón (Madrid) Spain.

[2] Dpt. Animal and Crop Science, University College Dublin, Dublin, Ireland.

in target environments, and selecting superior individuals to improve the frequency of favorable alleles (Falconer et al., 1996). The selection process of plant breeding has undergone three major transformations.

- Since the Neolithic revolution 10,000 years ago, the selection was based on the morphology and the appearance of traits. Traits such as non-shattering, grain size, or taste were the main selections during the domestication process. This selection is called **Phenotypic selection** and it has been and still is a crucial step in breeding programs. Nevertheless, the main drawback in phenotypic selection is the high evaluation cost of progenies and the time required for it.
- The development of the polymerase chain reaction by Kary Mullis allowed introducing marker selection to the breeder's toolbox (**Marker assisted selection**). In this sense, molecular marker technology has been used since the 1980s to help plant breeders to make a selection. With this tool breeders could indirectly by the association between phenotypes and genetic markers physically linked to causal loci, select individuals in early stages and with more efficiency than phenotypic selection. Marker-assisted selection efficiency is optimal when the genetic architecture of the traits under selection depends on a small number of genes with large effects. With molecular marker technology, a limited fraction of the genetic variation is explained by the identification of quantitative trait loci (QTL), and therefore only a limited proportion of genetic variance can be captured by the markers.
- With the introduction of the next-generation sequencing in 2007, genotyping became cheaper and breeders expanded the marker technology towards the entire genome. This selection known as genomic selection (GS) can estimate breeding values for quantitative traits based on whole-genome individuals through the simultaneous estimation of marker effects (Bernardo, 1994; Meuwissen et al., 2001).

This set of tools that studies whole genomes by integrating multiple disciplines with new technology from informatics and robotic systems to improve selection and mating in plant breeding programs is called **Genomic Assisted breeding (GAB)**. In GAB, other tools such as genetic transformation and genome editing will play a key role in the next few decades to select better-adapted individuals while pursuing faster genetic gains.

The rate of genetic gain must be enhanced to meet humanity's demand for agricultural products in the next few decades (Xu et al., 2020). In this regard, GS has been considered most promising for genetic improvement of complex traits controlled by many genes each with minor effects because (i) GS can increase the rates of genetic gain through increased accuracy of estimated breeding values, (ii) significantly shorter breeding cycles, and (iii) better utilization of available genetic resources through genome-guided mate selection (Akdemir and Sánchez, 2016). Therefore, breeding strategies that combine the power of GS and the potential of an extensive collection of germplasms, assisted by new technologies, will offer promise in crop breeding to contribute to global food security.

The use of genomic information for prediction was proposed by Bernardo (1994) but Meuwissen et al. (2001) proposed the current methodology to deal with the multicollinearity problem under genomic studies. Genomic selection (GS) uses supervised learning for predicting genomic estimated breeding values (GEBVs) of un-phenotyped individuals by using genome-wide molecular markers. Genomic prediction models are built using training data, i.e., genomic and phenotypic data for a set of individuals that is used to estimate marker (or lines) effects. The selection of individuals is based on the GEBVs and the performance of the GS model is determined by calculating the correlation between GEBVs and true breeding values. Enhancing GS accuracy is very important for the success of GS breeding programs, since the expected genetic gain from GS is directly proportional to the accuracy of GS models. The main factors affecting the accuracy in GS (Zhong et al., 2009; Liu et al., 2018; Zhang et al., 2019; Isidro et al., 2016) are heritability, linkage disequilibrium, population size, marker density and types, the relationship between TRS and TS, and statistical models. These factors interact in a complex relationship network among them (Isidro et al., 2016).

Training population optimization for genomic selection

The use of GS in breeding programs is potentially costly without a careful design of populations. The selection of the TRS is critical (Clark et al., 2012; Albrecht et al., 2011) for the success of GS since the predictions are based on markers or individual effects estimated on the TRS. If these estimates are wrong, the viability and effectiveness of the model performance will be null. The selection of the optimal TRS from a large population is called TRS optimization and consists of selecting the best training individuals that best predict the test set population. In the last few years, there has been a great interest in this research topic, mainly because TRS optimization can reduce phenotyping which is the current bottleneck in plant breeding programs. In addition, the traditional random selection of the TRS, or the selection based on breeder decisions, not always implied high predictability.

The design of the TRS was initially started with animal breeding (Habier et al., 2007; 2010; Clark et al., 2012; Pszczola et al., 2012). These studies and the subsequent plant breeding ones (Windhausen et al., 2012; Wientjes et al., 2013) focused on the relationship between TRS and TS. In the last decade, several studies examined the importance of optimization of the TRS by comparing specific selection criteria to random sampling.

The importance of using statistical approaches to develop an optimal TRS was first shown by Rincent et al. (2012). In this study, Rincent et al. (2012) defined which individuals were the optimal ones from a calibration set to predict a test set of candidates. Based on concepts from the mixed model equations introduced by Laloë (1993), they introduced criteria that aimed to maximize the reliability (coefficient of determination (CD), the square correlation between GEBVs and true breeding values) or minimized the prediction error variance (PEV) on the calibration set. In this study, they used a generalized version of CD and PEV (contrast between breeding values). They showed that the optimization criteria improved prediction accuracy when comparing with random sampling. To study the role of population structure in TRS optimization, Isidro et al. (2015) proposed stratified sampling and stratified CD as alternative algorithms to improve the optimization of TRS under population structure. In this study, authors concluded that the TRS optimization depended on the interaction of trait architecture and the ability of the criteria to capture phenotypic variance. For instance, when the population structure effects were highly stratified random sampling showed greater accuracies than CD or PEV criteria. In the same year, Akdemir et al. (2015) derived a computationally efficient approximation to the PEV based on principal components of the individuals as a criterion for TRS design, which showed less computational burden than previous criteria. These papers opened the door to other strategies for TRS optimization.

Bustos-Korts et al. (2016) proposed a TRS construction method that uniformly sampled the genetic space comprised by the target population of individuals, although, the results were similar to CD-mean. Other studies also stressed the importance of considering another way to construct the TRS by random sampling (Lorenz and Smith, 2015; He et al., 2016; Neyhart et al., 2017; Cericola et al., 2017; Norman et al., 2018; Olatoye et al., 2020; de Bem Oliveira et al., 2020), clustering approaches (Isidro et al., 2015; Akdemir et al., 2015; Bustos-Korts et al., 2016; Rincent et al., 2017; Norman et al., 2018; Guo et al., 2019; Sarinelli et al., 2019; Adeyemo et al., 2020), by using different levels of relatedness between TRS and TS (Lorenz and Smith, 2015; Roth et al., 2020) or by using other alternative algorithms to CD-mean and PEV-mean such as the different design matrix algorithm (Akdemir and Isidro-Sánchez, 2019), estimated theoretical accuracy (EthAcc) (Mangin et al., 2019), upper bound reliability (Yu et al., 2020) or the Fast and Unique Representative Subset Selection (FURS) (Guo et al., 2014). A criterion that is derived directly from Pearson's correlation between GEBVs and phenotypic values of the test set derived from the GBLUP model showed higher predictive ability than CD and PEV (Ou and Liao, 2019). Most aforementioned approaches above, do not use information from the test set (TS) while building the TRS. This was addressed by Lorenz and Smith (2015); Akdemir and Isidro-Sánchez (2019) and Ou and Liao (2019), showing that the information about the TS individuals when building the TRS leads to significant increases in accuracies.

In the context of hybrid breeding, the TRS optimization also plays an important role in comparing TRS optimization with traditional crossing design methods. In the hybrid context, all potential hybrids can be derived from the inbred individuals. In this sense, GS could be used for predicting general combining ability of lines whether or not they were evaluated as a hybrid, or directly to predict the hybrid performance to select and advance the most promising individuals. Genomic hybrid studies have been mainly performed in maize Genomic models have been applied to hybrid prediction mainly in maize (Bernardo, 1994; Schrag et al., 2009; Technow et al., 2014; Marulanda et al., 2016; Fristche-Neto et al., 2018; Seye et al., 2020; Kadam et al., 2016; Fritsche-Neto et al., 2021), in wheat (Zhao et al., 2013; 2014; Longin et al., 2015; Zhao et al., 2015; Marulanda et al., 2016; Schulthess et al., 2017), and less in other crops such as rye (Wang et al., 2014), or sunflower (Reif et al., 2013; Mangin et al., 2017; Dimitrijevic and Horn, 2018; Heslot and Feoktistov, 2020). These studies have emphasized the interest in using TRS optimization compared to the traditional crossing designs.

In general, there is no universal criterion to perform optimally under all the TRS optimization scenarios. It will depend highly on the factors affecting the QTL traits, mainly on heritability (Hayes et al., 2009), linkage disequilibrium between markers on TRS vs. TS, relationship between TRS and TS (Habier et al., 2007; Goddard, 2009), genetic architecture of the trait (McClellan et al., 2007; Jannink, 2010; Burstin et al., 2015), and population structure (Isidro et al., 2015).

Software for training set optimization

In addition, the number of software tools for implementation of TRS optimization is limited. There are three main softwares developed in R for public use:

1. The package STPGA Akdemir (2017) is a R package that uses a modified GA for solving subset selection problems but also allows users to chose from many predefined or user-defined criteria.
2. The package TSDFGS Ou and Liao (2019) is a R package that focuses on optimization of the TRS by a genetic algorithm (GA) and can be used for TRS optimization based on three built-in design criteria.

3. The package TrainSel Akdemir et al. (2021) designed a new package to provide many more options than previous softwares, for example, the ability to select multiple sets from multiple candidate sets, specification of whether or not the resulting set needs to be ordered, or the power to perform multi-objective optimization.

In the following section, we provided a practical guideline to perform TRS optimization analysis using R.

Hands on training optimization using TrainSel package in R

In this section, we will illustrate the use of the package 'TrainSel'. We will use the data sets provided within the package under the object named 'WheatData' throughout this presentation. The original data was obtained from the webpage https://triticeaetoolbox.org/. The data contains the genomewide marker data (at 4670 markers) and simulated trait data (phenotypic measurements are simulated using an infinitesimal model) for 200 wheat varieties.

We can install, load the library and then load the data 'WheatData' using the following code:

```
library(devtools)
install_github("TheRocinante-lab/TrainSel")
library("TrainSel")
data("WheatData")
```

We can load the library and 'WheatData' using the following code:

```
library(TrainSel)
## Loading required package: cluster
data("WheatData")
```

Marker data is contained in the matrix object 'Wheat.M', a relationship matrix for the individuals calculated from the marker matrix in 'Wheat.K', and the plant height measurements are in 'Wheat.Y'. We can see the format of this data:

```
Wheat.M[1:5,1:5]
##          IWA1   IWA2   IWA3   IWA4   IWA5
## Line1     1      1     -1     -1     -1
## Line2     1      1      1     -1     -1
## Line3     1      1      1     -1     -1
## Line4     1      1      1      1     -1
## Line5     1      1      1     -1     -1
```

```
Wheat.K[1:5,1:5]
##              Line1      Line2      Line3      Line4      Line5
##Line1  1.9578361  0.8261588  0.7546713  0.5149389 -0.2697654
##Line2  0.8261588  2.0033376  0.5756170  0.5241991 -0.2315324
##Line3  0.7546713  0.5756170  1.9807687  0.7077443 -0.2888690
##Line4  0.5149389  0.5241991  0.7077443  2.2816612 -0.2645939
##Line5 -0.2697654 -0.2315324 -0.2888690 -0.2645939  1.8086996
Wheat.Y[1:5,]
##          id plant.height
## 1846 Line1     138.4471
## 250  Line2     122.5955
## 1541 Line3     134.7883
## 1516 Line4     121.4741
## 508  Line5     127.3920
```

Selection of a subset of individuals for a phenotypic experiment in a single environment

We aim to build a predictive model for the plant height based on the genome-wide marker data. The dataset contains the plant heights for all these individuals. However, assume that we do not have the plant heights measurements for these 200 individuals and currently only a subset of 160 candidate individuals is available for use in the phenotypic experiment. Furthermore, since measuring plant height for 160 candidate individuals via a phenotypic experiment can be costly, we assume that we can only perform the phenotypic experiment with say a maximum of 50 training individuals selected from the 160. Following this phenotypic experiment, we can use the available marker data and the measured height of these 50 individuals to train a genomic prediction model to make inferences about the heights of the remaining 110 individuals or any other set of individuals that have the same genome-wide marker data, for example, the 40 individuals that were not available for the phenotypic experiment.

In the following subsections, we explain the different TRS optimization scenarios using TrainSel. These scenarios cover model-based and non-parametric approaches, targeted-untargeted optimization, use of environmental covariates for obtaining a design along with a TRS, and design of multi-environmental GS-experiments where the genomic co-variances of the genetic effect among the environments is utilized. The examples below are listed in an increasing order of complexity. The aim is to present certain capabilities of the package so that the users can adopt these examples to their TRS complex optimization needs.

Selection of a subset of individuals for a homogeneous design in a single environment

The simplest case used for 'TrainSel' is the situation where the phenotypic experiment will only be performed in one homogeneous environment. In this case, our purpose is to select a subset of size 50 from the 160 available individuals so that the genomic prediction models that are trained on it have a good generalization performance.

We distinguish between two cases of optimal TRS selection based on whether, we seek to generalize well for a specific target set of individuals (Targeted optimization), or not (Untargeted optimization). Not all optimization criteria are sensitive to this distinction (D-opt, A-opt, PAM, mean genetic distance in TRS), however, when it is so (CDmean, PEVmean, mean genetic distance between TRS and TS, etc., ...) this is reflected in how the optimization criteria are calculated. Previous results show that optimization methods that include information from the test set (targeted) showed the highest accuracies, indicating that a priori information from the TS improves genomic predictions.

Untargeted optimization

Many different statistics can be used for the untargeted optimization with a homogeneous design in a single environment. The package 'TrainSel' is designed to be flexible for use with any design criteria set, however, this means that the users need to program their optimization functions. Below are some example functions that should get started for writing your own:

D-optimality criterion

D-optimality criterion is a model-based design criterion. The underlying model for the D-optimality criterion is a linear model. For the problem of selection of a subset of individuals for a homogeneous design in a single environment a linear model relating the genotypic data to phenotypic measurements in the training data can be expressed as

$$y = 1\mu + f(M)\beta + \epsilon,$$

where y is the n vector of phenotypic measurements in the training data, μ is a scalar parameter for the mean of the phenotypic measurements, M is the $n \times m$ marker matrix for the n training individuals, $f(M)$ is a genomic features matrix with dimensions $n \times q$, β is the q vector of genomic feature effects, and ϵ is the residual error vector of length n. We further assume that the elements of ϵ are independent and identically distributed with a normal distribution with zero mean and variance σ_e^2. Under this model the variance of the estimators for β is known to be proportional to $[f(M)'f(M)]^{-1}$. D-optimal selection of n training individuals from N individuals in the candidate set involves minimizing the determinant of this matrix (or equivalently maximizing the log-determinant of

$f(M)'f(M)$). The feature matrix $f(M)$ is usually the first q principal components matrix for the marker matrix M and for this measure to be defined q should be less than n.

```
#We will use the first 30 principal components for this
Wheat.M_centered<-scale(Wheat.M,center=TRUE,
scale=FALSE)
svdWheat.M_centered<-svd(Wheat.M_centered,nu=30,
nv=30)
PC<-Wheat.M_centered%*%svdWheat.M_centered$v
dim(PC)
## [1] 200  30
dataDopt<-list(FeatureMat=PC)
DOPT<-function(soln, Data){
Fmat<-Data[["FeatureMat"]]
  return(determinant(crossprod(Fmat[soln,]),
  logarithm=TRUE)$modulus)
}
```

In all the following examples, we will set the TrainSel algorithm parameters as follows: Number of iterations for the GA is 100, population size for GA is 30, and the number of elite solutions at each iteration is 3. Depending on the size and complexity of the optimization problem we would like to adjust these parameters.

```
TSC<-TrainSelControl()
TSC$niterations=100
TSC$npop=30
TSC$nelite=3
TSOUTD<-TrainSel(Data=dataDopt,
            Candidates = list(1:160),
            setsizes = c(50),
            settypes = "UOS",
            Stat = DOPT, control=TSC)


## [1] "Line3" "Line4" "Line7" "Line13" "Line15"
"Line16"
```

Convergence should be checked after the algorithm ends. We can do this by checking the path of the objective function through the iterations:

```
plot(TSOUTD$maxvec)
```

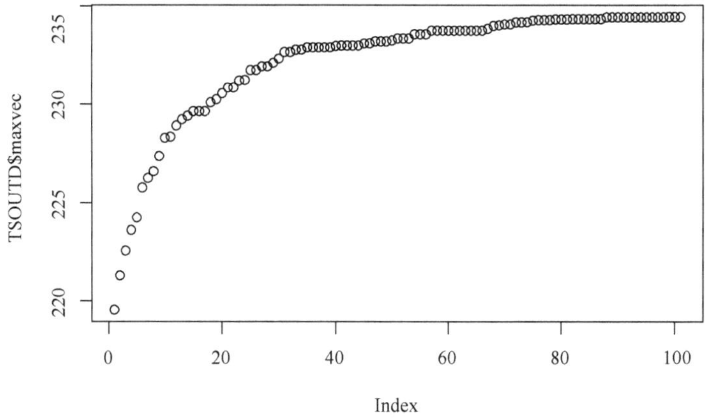

Fig. 1. The plot of the objective value path across the iterations. The flattening of this curve to a convergence point indicates local convergence, In this example, we see that increasing the number of iterations or the population size for the genetic algorithm will be needed.

CDMEAN-optimality criterion

CDMEAN-optimality criterion is also a model-based criterion based on a G-BLUP mixed model. For the problem of selection of a subset of individuals for a homogeneous design in a single environment, a G-BLUP model relating the genotypic data to phenotypic measurements in the training data can be expressed as

$$y = 1\mu + Zu + \epsilon$$

with μ a scalar parameter for the mean of the phenotypic measurements, Z the $n \times N$ design matrix for the N individuals in the candidate set, $\epsilon \sim N_n(0, \sigma_e^2)$ independent of $u \sim N_q(0; \sigma_q^2 G)$.
For this model, the coefficient of determination matrix of $\hat{u}$ for predicting u is given by

$$(GZ' P ZG) \oslash G$$

where $P = V^{-1} - V^{-1}1(1'V^{-1}1)^{-1}1'V^{-1}$ is the projection matrix and $\oslash$ expresses the element-wise division.

The diagonals of this matrix are the coefficients of determination of the predictions for individual individuals and the mean of these coefficients of determination values over the selected individuals is called the CDMEAN-optimality criterion.[1] CDMEAN criterion takes values between 0 and 1 and the larger values are preferable. For the untargeted optimization, the usual practice is to use CDMEAN that is calculated over the individuals not

[1] A risk-averse approach would entail maximizing the minimum of selected diagonals. Another approach involves the calculation of the CD matrix for a given set of contrasts then taking the mean of the diagonals of this matrix (Rincent et al., 2012; 2017).

included in the training set. We can program this objective function to be used in 'TrainSel' as follows:

```
# note that we are not using the target individuals
dataCDMEANopt<-list(G=Wheat.K[1:160,1:160],
lambda=1)
CDMEANOPT<-function(soln, Data){
 G<-Data[["G"]]
 lambda<-Data[["lambda"]]
 Vinv<-solve(G[soln,soln]+lambda*diag(length(soln)))
 outmat<-(G[,soln]%*%(Vinv-(Vinv%*%Vinv)/
 sum(Vinv))%*%G[soln,])/G
 return(mean(diag(outmat[-soln,-soln])))
}
TSOUTCD<-TrainSel(Data=dataCDMEANopt,
           Candidates = list(1:160),
           setsizes = c(50),
           settypes = "UOS",
           Stat = CDMEANOPT, control=TSC)
head(rownames(Wheat.M)[TSOUTCD$BestSol_int])
## [1] "Line6" "Line8" "Line11" "Line15" "Line17"
"Line18"
```

Maximin distance criterion

Maximin distance criterion is a non-parametric design criterion. An optimal training set of size n from the N candidates is selected by maximizing the minimum[2] genetic distance among the training individuals, so this is a space-filling design. Next, we show how to program this criterion in R:

```
dataMaximin<-list(DistMat=as.matrix(dist(Wheat.M_
centered)))
MaximinOPT<-function(soln, Data){
  Dsoln<-Data[["DistMat"]][soln,soln]
  DsolnVec<-Dsoln[lower.tri(Dsoln,diag=FALSE)]
  return(min(DsolnVec))
}
TSOUTMaximin<-TrainSel(Data=dataMaximin,
            Candidates = list(1:160),
            setsizes = c(50),
            settypes = "UOS",
            Stat = MaximinOPT, control=TSC)
```

[2] We could also maximize the mean distance leading to Maximean criterion.

```
head(rownames(Wheat.M)[TSOUTMaximin$BestSol_int])
## [1] "Line3" "Line5" "Line9" "Line11" "Line15"
"Line16"
```

Targeted optimization

When the focus is on making inferences about the trait values for a known target set of individuals using genomic prediction, we can use what we call targeted optimization criteria.

Mean PEV criterion based on linear model

Dopt criterion is not sensitive to information about the target set of individuals. Nevertheless, a related linear model-based criteria called the mean prediction error variance (Mean PEV) can be used when the genotypic data for the target set is available. This criterion relates to the average prediction error variance of the predictions for the target set of individuals using the linear model in Equation 1.

```
##Target individuals are in PC
dataPEVlm<-list(FeatureMat=PC, Target=161:200)
PEVlmOPT<-function(soln, Data){
  Fmat<-Data[["FeatureMat"]]
  targ<-Data[["Target"]]
  return(mean(diag(Fmat[targ,]%*%solve
  (crossprod(Fmat[soln, ]))%*%t(Fmat[targ,]))))
}
TSOUTPEVlm<-TrainSel(Data=dataPEVlm,
          Candidates = list(1:160),
          setsizes = c(50),
          settypes = "UOS",
          Stat = PEVlmOPT, control=TSC)
 head(rownames(Wheat.M)[TSOUTPEVlm$BestSol_int])
## [1] "Line1" "Line2" "Line3" "Line7" "Line8"
"Line13"
```

Targeted CDMEAN criterion

```
##This time the Target individuals are in G
dataCDMEANTargetOpt<-list(G=Wheat.K, lambda=1,
Target=161:200)
```

```
CDMEANOPTTarget<-function(soln, Data){
 G<-Data[["G"]]
 lambda<-Data[["lambda"]]
 targ<-Data[["Target"]]
 E<-matrix(1,nrow=length(soln),ncol=1)
 Vinv<-solve(G[soln,soln]+lambda*diag(length(soln)))
 outmat<-(G[,soln]%*%(Vinv-Vinv%*%E%*%solve(t(E)%*
 %Vinv%*%E)%*%t(E)%*%Vinv)%*%G[soln,])/G
 return(mean(diag(outmat[targ,targ])))
}
TSOUTCDTarg<-TrainSel(Data=dataCDMEANTargetOpt,
           Candidates = list(1:160),
           setsizes = c(50),
           settypes = "UOS",
           Stat = CDMEANOPTTarget, control=TSC)
head(rownames(Wheat.M)[TSOUTCDTarg$BestSol_int])
## [1] "Line6"   "Line8"    "Line9"    "Line10"
"Line13" "Line14"
```

Multiple Design Criterion

Maximize the mean genomic distance of training individuals and simultaneously minimize the mean distance to the target individuals.

```
dataMultOpt<-list(DistMat=as.matrix(dist(Wheat.M_
centered)))
MultOPT<-function(soln, Data){
   D<-Data[["DistMat"]]
   Dsoln<-D[soln,161:200]
   DsolnVec1<- -mean(c(unlist(Dsoln)))
   Dsoln2<-D[soln,soln]
   DsolnVec2<-mean(c(unlist(Dsoln2)))
   return(c(DsolnVec2,DsolnVec1))
}
TSOUTMultOPt<-TrainSel(Data=dataMultOpt,
            Candidates = list(1:160),
            setsizes = c(50),
            settypes = "UOS",
            Stat = MultOPT, nStat=2, control=TSC)
```

Plot the frontier solutions:

```
FrontierSols<-t(TSOUTMultOPt$BestVal)
plot(FrontierSols, xlab="mean dist in train",
     ylab="-mean dist to target")
```

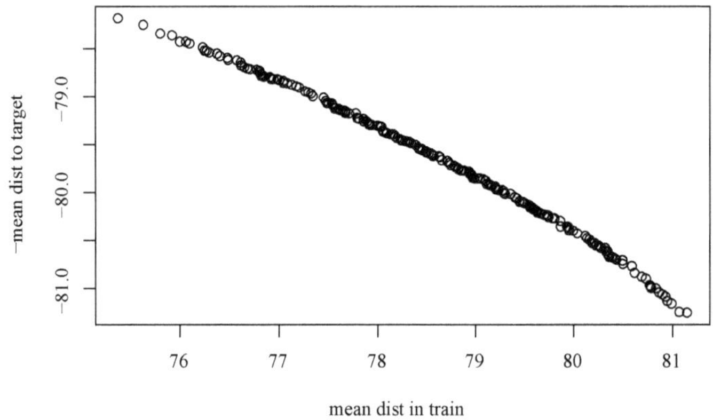

Fig. 2. The frontier curve obtained by the TrauinSel algorithm for maximizing the mean genetic distance in the training set and also to minimize the mean genetic distance of the training set and the target set (maximize the negative of the mean genetic distance of the training set and the target set). The solutions to the left most are training sets that are closest to the target and on the other hand the solutions to the right are more diverse. The solutions located at the center are trade-off training sets that balance these two objectives.

Note that we have many solutions on the frontier curve. We get solutions that give a good balance by selecting a few of these by constraining the values of the objectives. Then we display the first of these selected solutions on a principal components plot.

```
Solns<-which((FrontierSols[,1]>=79) & (FrontierSols
[,2]>= -80))# subset of solutions
Selected<-TSOUTMultOPt$BestSol_int[, Solns[1]]
Groups<-rep("Cand",200)
Groups[161:200]<-"Targ"
Groups[Selected]<-"TRS"
PlotData<-data.frame(x1=PC[,1], x2=PC[,2], Groups=
factor(Groups))
library(ggplot2)
p<-ggplot(PlotData, aes(x=x1,y=x2, color=Groups))
+geom_point()
p
```

Selection of a subset of individuals for a known design in a single environment

In certain cases, we are looking for conducting a phenotypic experiment with n training individuals selected out of N candidate individuals but in addition, we also have a particular blocking structure and environmental covariates involved in the design of the experiment. Suppose the matrix E is the $n \times p$

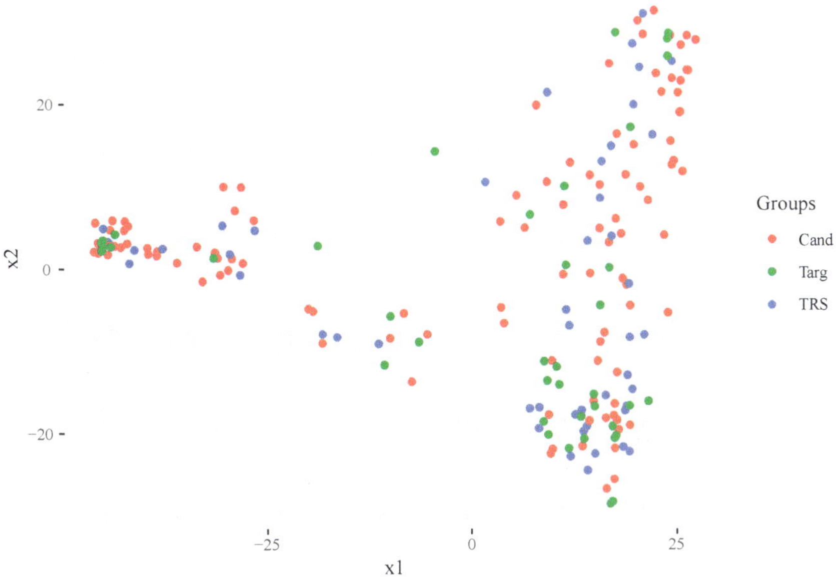

Fig. 3. The representation of a trade-off solution that was selected from the central part of the frontier curve in Figure 2 on the principal components of the genomic relationship matrix for a selected set of candidate and target genotypes. The selected training set is shown with blue dots, remainder set with red, target set with green dots.

environmental covariates matrix. For instance, this matrix could be the design matrix for a row-column blocking within the environment. We assume still that we want to make inferences about the genomic values after accounting for these covariates. In this case, the order in which the individuals are positioned in the environment will be important. Perhaps, we would like to use similar individuals in dissimilar blocks, and also we would like to observe genetically distant individuals within similar blocks. We refer to this kind of optimization as the "ordered" optimization as opposed to the "unordered" optimization.

Untargeted optimization

D-optimality criterion with environmental covariates

D-optimality criterion can be easily adopted for this purpose. First, we write the model as,

$$y = E\beta_{env} + f(M)\beta_f + \epsilon$$

where y is the n vector of phenotypic measurements in the training data, E is the $n \times p$ design matrix for the environmental covariates, β_{env} is the p vector of the effects of the environmental covariates, M is the $n \times m$ the marker matrix for the n training individuals, $f(M)$ is a genomic features matrix with dimensions $n \times q$, β_f is the q vector of effects of genomic features, and ϵ is the residual error vector of length n. We further assume that the elements of ϵ are independent and identically distributed with a normal distribution with zero mean and variance σ_e^2. Under this model the variance of the estimators for β is known to be proportional to $[f(M)'(I - E(E'E)^{-1}E')f(M)]^{-1}$. D-optimal selection of n training individuals from N individuals in the candidate set involves minimizing the determinant of this matrix (or equivalently maximizing the log-determinant of $f(M)'(I - E(E'E)^{-1}E')f(M)$). The matrix $(I - E(E'E)^{-1}E')$ is the projection matrix to the orthogonal space of the column space of E.

```
E<-data.frame(expand.grid(row=paste("row",1:5,
                                    sep="_"),
                          col=paste("col",1:10,
                                    sep="_")))
E$row<-as.factor(E$row)
E$col<-as.factor(E$col)
DesignE<-model.matrix(~row+col+row*col, data=E)
P<-diag(nrow(DesignE))-DesignE%*%solve(crossprod(D
esignE))%*%t(DesignE)
dataDoptEnv<-list(FeatureMat=PC, Projection=P)
DOPTwithE<-function(soln, Data){
  Fmat<-Data[["FeatureMat"]]
  P<-Data[["Projection"]]
  return(determinant(crossprod(P%*%Fmat[soln,]),
  logarithm=TRUE)$modulus)
}
TSOUTDwithE<-TrainSel(Data=dataDoptEnv,
             Candidates = list(1:160),
             setsizes = c(50),
             settypes = "OS",
             Stat = DOPTwithE, control=TSC)
TSOUTDwithE$BestSol_int #order of this is important
## [1]   7 105  54  97  12 101 108 133 147 109  26  51
87  61 148  69  71  72 124
##[20]  21  20  13  40 140 117  16 149  42 153  30  45
15  18  90  75 158   4  47
##[39]  17 128  41 146 139 132 144 112 135  62 121 125
head(rownames(Wheat.M)[TSOUTDwithE$BestSol_int])
## [1] "Line7" "Line105" "Line54" "Line97" "Line12"
"Line101"
```

```
E$GID<-rownames(Wheat.M)[TSOUTDwithE$BestSol_int]
###Here is the final design
head(E)
```

```
##       row     col      GID
## 1  row_1   col_1    Line7
## 2  row_2   col_1  Line105
## 3  row_3   col_1   Line54
## 4  row_4   col_1   Line97
## 5  row_5   col_1   Line12
## 6  row_1   col_2  Line101
```

CDMEAN-optimality criterion with environmental covariates

We can also use environmental covariates with the CDMEAN-optimality criterion. In order to do this we first need to add the environmental covariates into the G-BLUP model. This model is written as,

$$y = E\beta_{env} + Zu + \epsilon$$

with E is the $n \times p$ design matrix for the environmental covariates, β_{env} is the p vector of the effects of the environmental covariates, Z is the $n \times N$ design matrix for the N individuals in the candidate set, $\epsilon \sim N_n(0, \sigma_e^2 I)$ is independent of $u \sim N_q(0; \sigma_g^2 G)$.

For this model, the coefficient of determination matrix of $\hat{u}$ for predicting u is given by

$$(GZ' P ZG) \oslash G$$

where $P = V^{-1} - V^{-1}E(E'V^{-1}E)^{-1}E'V^{-1}$ is the projection matrix and $\oslash$ expresses the element-wise division.

```
dataCDMEANoptwithEnv<-list(G=Wheat.K[1:160,1:160],
E=DesignE, lambda=1)
CDMEANOPTwithEnv<-function(soln, Data){
 G<-Data[["G"]]
 E<-Data[["E"]]
 lambda<-Data[["lambda"]]
 Vinv<-solve(G[soln,soln]+lambda*diag(length(soln)))
 outmat<-(G[,soln]%*%
     (Vinv-Vinv%*%E%*%solve(t(E)%*%Vinv%*%E)%*%t(E)
     %*%Vinv)%*%G[soln,])/G
 return(mean(diag(outmat[-soln,-soln])))
}
TSOUTCDwithENV<-TrainSel(Data=dataCDMEANoptwithEnv,
               Candidates = list(1:160),
               setsizes = c(50),
               settypes = "OS",
               Stat = CDMEANOPTwithEnv, control=TSC)
```

```
E$GID<-rownames(Wheat.M)[TSOUTCDwithENV$BestSol_int]
###Here is the final design
head(E)
```

```
##        row    col      GID
## 1    row_1  col_1  Line111
## 2    row_2  col_1   Line48
## 3    row_3  col_1   Line53
## 4    row_4  col_1   Line62
## 5    row_5  col_1  Line119
## 6    row_1  col_2   Line32
```

Targeted optimization and allowing for replicates

We can also do targeted optimization in this case with minimal change to the last code:

```
dataCDMEANoptwithEnvTarget<-list(G=Wheat.K,
E=DesignE,Target=161:200, lambda=1)
CDMEANOPTwithEnvTarget<-function(soln, Data){
 G<-Data[["G"]]
 E<-Data[["E"]]
 targ<-Data[["Target"]]
 lambda<-Data[["lambda"]]
 Vinv<-solve(G[soln,soln]+lambda*diag(length(soln)))
 outmat<-(G[,soln]%*%
          (Vinv-Vinv%*%E%*%solve(t(E)%*%Vinv%*%E)%
          *%t(E)%*%Vinv)
          %*%G[soln,])/G
 return(mean(diag(outmat[targ,targ])))
}
TSOUTCDwithENVTarg<-TrainSel(Data=dataCDMEANoptwit
hEnvTarget,
        Candidates = list(1:160),
        setsizes = c(50),
        settypes = "OMS",
        Stat = CDMEANOPTwithEnvTarget, control=TSC)
E$GID<-rownames(Wheat.M)
[TSOUTCDwithENVTarg$BestSol_int]
###Here is the final design
head(E)
```

```
##        row    col      GID
## 1  row_1  col_1  Line111
## 2  row_2  col_1   Line68
## 3  row_3  col_1  Line153
## 4  row_4  col_1   Line24
## 5  row_5  col_1   Line55
## 6  row_1  col_2   Line82
```

Design for a phenotypic experiment in multiple environments

In practice, most genomic selection experiments are performed over multiple environments. Designing genomic selection over multiple environments means we would like to choose training individuals for use in each of these environments and for genomic selection the distribution of the alleles within and between the different environments can be arranged optimally for obtaining better generalization performance.

Using CDMEAN-optimality criterion for genomic selection experiment design in multiple environments

As before, we first need to state the underlying model. For multi-environmental trials, a commonly used genomic prediction model is the multi-environmental G-BLUP model. Suppose we have 3 environments, in Environment 1 we can accommodate 30 individuals in a 6-rows 5-columns design, in Environment 2 we can accommodate 20 individuals in a 4-rows 5-columns design, and in Environment 3 we can accommodate 50 individuals in an unknown homogeneous design. We assume that the genomic covariance of the environments is (proportionally) equal to the matrix

$$V_g = \begin{pmatrix} 1.0 & .7 & .5 \\ .7 & 1.2 & .8 \\ .5 & .8 & 1.5 \end{pmatrix}$$

We assume that the residual errors in these environments are correlated with the following covariance matrix:

$$V_e = \begin{pmatrix} 1.0 & 0 & 0 \\ 0 & 1.5 & 0 \\ 0 & 0 & 1.0 \end{pmatrix}$$

We can express the model for the training data as follows:

$$\begin{pmatrix} y_1 \\ y_2 \\ y_3 \end{pmatrix} = E\beta_{env} + \begin{pmatrix} Z_1u_1 \\ Z_2u_2 \\ Z_3u_3 \end{pmatrix} + \begin{pmatrix} Z_1\epsilon_1 \\ Z_2\epsilon_2 \\ Z_3\epsilon_3 \end{pmatrix}$$

where E is the $n \times p$ design matrix for the environmental covariates, β_{env} is the p vector of the effects of the environmental covariates, y_i is a n_i vector and Z_i is the $n_i \times N$ design matrix for the N individuals in the candidate set in environment i for $i = 1, 2, 3$. In addition, we assume $(\epsilon_1, \epsilon_2, \epsilon_3) \sim N_{N\times3}(0, I_N, V_e)$ is independent of $(u_1, u_2, u_3) \sim N_{N\times3}(0; G, V_g)$. This means that the vectorized form of these matrices $\epsilon = vec(\epsilon_1, \epsilon_2, \epsilon_3)$ and $u = vec(u_1, u_2, u_3)$ are independently distributed as $N_{3N}(0, V_e \otimes I_N)$ and $N_{3N}(0, V_g \otimes G)$.

In our case $n_1 = 30$, $n_2 = 20$, $n_3 = 50$, and $n = n_1 + n_2 + n_3 = 100$. Here is how you can approach this problem using 'TrainSel':

```
Vg=matrix(c(1.0, .7, .5, .7, 1.2, .8, .5, .8, 1.5),
3,3)
Ve=matrix(c(1.0, 0,0, 0,1.5,0, 0,0,1.0), 3,3)
rownames(Vg)<-colnames(Vg)<-rownames(Ve)<-
colnames(Ve)<-paste("E",1:3, sep="")
G<-kronecker(Vg, Wheat.K, make.dimnames = TRUE)
R<-kronecker(Ve, diag(nrow(Wheat.K)), make.dimnames
= TRUE)
#### Note the shape of G (same as R)
head(rownames(G))
## [1] "E1:Line1" "E1:Line2" "E1:Line3" "E1:Line4"
"E1:Line5" "E1:Line6"
 tail(rownames(G))
## [1] "E3:Line195" "E3:Line196" "E3:Line197"
"E3:Line198" "E3:Line199"
## [6] "E3:Line200"
 dim(G)
## [1] 600 600
```

By examining the shape of the G matrix above we see that the candidate set is on the 1st through 160th, 200th through 360th and, 400th through 560th rows and columns of this matrix. We want to use 30 from 1 through 160, 20 from 200 through 360, and 50 from 400 through 560. The first two sets are ordered and the last one is unordered. We are also going to assume duplicates within an environment are not allowed.

The design Matrix for the environments is obtained below. I am assuming no interaction effects between rows and columns.

```
E1<-(expand.grid(row=paste("row",1:6, sep="_"),
                 col=paste("col",1:5, sep="_")))
E2<-(expand.grid(row=paste("row",1:4, sep="_"),
                 col=paste("col",1:5, sep="_")))
E3<-data.frame(row=paste("row",rep(1,50),sep="_"),
               col=paste("col",rep(1,50), sep="_"))
EnvData<-data.frame(Env=c(rep("E1", 30), rep("E2",
               20),rep("E3", 50)),
               rbind(E1,E2,E3))
DesignE<-model.matrix(~Env+Env/(row+col),
data=EnvData)
DesignE<-DesignE[,colSums(DesignE)>0]
## This is the names of the fixed effects design matrix
colnames(DesignE)
##[1]"(Intercept)"  "EnvE2"  "EnvE3"  "EnvE1:rowrow_2"
##[5] "EnvE2:rowrow_2" "EnvE1:rowrow_3" "EnvE2:rowrow_3"
"EnvE1:rowrow_4"
##[9] "EnvE2:rowrow_4" "EnvE1:rowrow_5" "EnvE1:rowrow_6"
"EnvE1:colcol_2"
##[13] "EnvE2:colcol_2" "EnvE1:colcol_3" "EnvE2:colcol_3"
"EnvE1:colcol_4"
##[17] "EnvE2:colcol_4" "EnvE1:colcol_5" "EnvE2:colcol_5"
dataCDMEANoptwithEnvME<-list(G=G,R=R, E=DesignE)
Target<-c(161:200, 361:400,561:600)
# we will exclude these since we assume nontargeted
CDMEANOPTwithEnvME<-function(soln, Data){
 G<-Data[["G"]]
 R<-Data[["R"]]
 E<-Data[["E"]]
 Vinv<-solve(G[soln,soln]+R[soln,soln])
 outmat<-(G[,soln]
          %*%(Vinv-Vinv%*%E%*%solve(t(E)%*%Vinv%*
          %E)%*%t(E)%*%Vinv)
          %*%G[soln,])/G
return(mean(diag(outmat[-c(soln,  Target),-c(soln,
Target)])))
#returning the mean CD on the remaining GIDs
#(1:200 after deleting Target and the training.)
}
```

```
TSOUTCDwithENVME<-TrainSel(Data=dataCDMEANoptwithEnvME,
         Candidates = list(1:160, 201:360,401:560),
         setsizes = c(30,20,50),
         settypes = c("OS","OS","UOS"),
         Stat = CDMEANOPTwithEnvME, control=TSC)
##Putting GIDs in the design #first 30 are Env1
E1$GID<-rownames(G)[TSOUTCDwithENVME$BestSol_
int[1:30]]
###Here is the final design for env 1
head(E1)
##       row    col        GID
## 1  row_1  col_1  E1:Line30
## 2  row_2  col_1   E1:Line3
## 3  row_3  col_1  E1:Line21
## 4  row_4  col_1   E1:Line6
## 5  row_5  col_1  E1:Line41
## 6  row_6  col_1 E1:Line126
##second 20 are Env2
E2$GID<-rownames(G)[TSOUTCDwithENVME$BestSol_
int[31:50]]
###Here is the final design for env 2
head(E2)
##       row    col        GID
## 1  row_1  col_1 E2:Line128
## 2  row_2  col_1 E2:Line109
## 3  row_3  col_1 E2:Line118
## 4  row_4  col_1  E2:Line87
## 5  row_1  col_2  E2:Line55
## 6  row_2  col_2  E2:Line93
##Last 50 are Env3
E3$GID<-rownames(G)[TSOUTCDwithENVME$BestSol_
int[51:100]]
###Here is the final design for env 3
head(E3)
## row colGID
## 1  row_1  col_1   E3:Line4
## 2  row_1  col_1   E3:Line5
## 3  row_1  col_1   E3:Line7
## 4  row_1  col_1   E3:Line8
## 5  row_1  col_1   E3:Line9
## 6  row_1  col_1  E3:Line10
```

Implementing a penalty function for the total number of individuals in the experiment

Suppose we want to restrict the total number of unique individuals used in the above multi-environmental design to between 80 and 90. We can use a penalty function approach to impose this constraint: A penalty function forces the optimization algorithm to find desired solutions by assigning the points that don't satisfy the values of the constraints that are small with respect to the values of the optimization criteria (the size of this value might depend on how far the solutions are from satisfying the constrains), if the constraints are satisfied then the penalty functions return zero.

```
PenaltyFunction<-function(soln){
 soln1<-soln[1:30]
 soln2<-soln[31:50]-200
 soln3<-soln[51:100]-400
 numuniquegeno<-length(unique(c(soln1,soln2,soln3)))
 if( (numuniquegeno<80) | (90<numuniquegeno)){
   return(min(numuniquegeno-80, 90-numuniquegeno))
     } else {
   return(0)
   }
}
CDMEANOPTwithEnvMEwithPenalty<-function(soln, Data){
  penalty<-PenaltyFunction(soln)
  if (penalty==0){
  G<-Data[["G"]]
  R<-Data[["R"]]
  E<-Data[["E"]]
  Vinv<-solve(G[soln,soln]+R[soln,soln])
  outmat<-(G[,soln]
           %*%(Vinv-Vinv%*%E%*%solve(t(E)%*%Vinv%*%
           E)%*%t(E)%*%Vinv)
           %*%G[soln,])/G
  return(mean(diag(outmat[-c(soln, Target),-c(soln,
  Target)])))
  } else {return(penalty)}
}
TSOUTCDwithENVMEwithPenalty<-TrainSel(Data=dataCDME
  ANoptwithEnvME,
  Candidates = list(1:160, 201:360,401:560),
  setsizes = c(30,20,50),
  settypes = c("OS","OS","UOS"),
  Stat = CDMEANOPTwithEnvMEwithPenalty, control=TSC)
```

```
E1$GID<-rownames(G)[TSOUTCDwithENVMEwithPenalty$BestSol_
int[1:30]]
###Here is the final design for env 1
head(E1)
```

```
##      row    col        GID
## 1  row_1  col_1  E1:Line95
## 2  row_2  col_1  E1:Line78
## 3  row_3  col_1  E1:Line29
## 4  row_4  col_1  E1:Line79
## 5  row_5  col_1  E1:Line55
## 6  row_6  col_1  E1:Line91
```

```
E2$GID<-rownames(G)[TSOUTCDwithENVMEwithPenalty$BestSol_
int[31:50]]
###Here is the final design for env 2
head(E2)
```

```
##      row    col         GID
## 1  row_1  col_1  E2:Line142
## 2  row_2  col_1   E2:Line53
## 3  row_3  col_1  E2:Line101
## 4  row_4  col_1   E2:Line99
## 5  row_1  col_2   E2:Line80
## 6  row_2  col_2   E2:Line46
```

```
E3$GID<-rownames(G)[TSOUTCDwithENVMEwithPenalty$BestSol_
int[51:100]]
###Here is the final design for env 3
head(E3)
```

```
##      row    col        GID
## 1  row_1  col_1   E3:Line4
## 2  row_1  col_1   E3:Line8
## 3  row_1  col_1  E3:Line11
## 4  row_1  col_1  E3:Line12
## 5  row_1  col_1  E3:Line15
## 6  row_1  col_1  E3:Line16
```

References

Adeyemo, E., Bajgain, P., Conley, E., Sallam, A. H. and Anderson, J. A. 2020. Optimizing training population size and content to improve prediction accuracy of fhb-related traits in wheat. *Agronomy* 10: 543.

Akdemir, D., Sanchez, J. I. and Jannink, J.-L. 2015. Optimization of genomic selection training populations with a genetic algorithm. *Genet. Sel. Evol.* 47: 38.

Akdemir, D. and Sánchez, J. I. 2016. Efficient breeding by genomic mating. *Frontiers in Genetics* 7.
Akdemir, D. (Ed.). 2017. STPGA: Selection of Training Populations by Genetic Algorithm. R package version 4.0.
Akdemir, D. and Isidro-Sánchez, J. 2019. Design of training populations for selective phenotyping in genomic prediction. *Scientific Reports* 9: 1–15.
Akdemir, D., Rio, S. and y Sánchez Julio, I. 2021. Trainsel: an r package for selection of training populations. *Frontiers in Genetics* 12: 607.
Albrecht, T., Wimmer, V., Auinger, H., Erbe, M., Knaak, C., Ouzunova, M., Simianer, H. and Schön, C. 2011. Genome-based prediction of testcross values in maize. *Theoretical and Applied Genetics* 12: 339–350.
Bernardo, R. 1994. Prediction of maize single-cross performance using rflps and information from related hybrids. *Crop Science* 34: 20–25.
Burstin, J., Salloignon, P., Chabert-Martinello, M., Magnin-Robert, J., Siol, M., Jacquin, F., Chauveau, A., Pont, C., Aubert, G., Delaitre, C. et al. 2015. Genetic diversity and trait genomic prediction in a pea diversity panel. *BMC Genomics* 16: 105.
Bustos-Korts, D., Malosetti, M., Chapman, S., Biddulph, B. and van Eeuwijk, F. 2016. Improvement of predictive ability by uniform coverage of the target genetic space. *G3: Genes| Genomes| Genetics* 6: 3733–3747.
Cericola, F., Jahoor, A., Orabi, J., Andersen, J. R., Janss, L. L. and Jensen, J. 2017. Optimizing training population size and genotyping strategy for genomic prediction using association study results and pedigree information. a case of study in advanced wheat breeding lines. *PloS One* 12: e0169606.
Clark, S. A., Hickey, J. M., Daetwyler, H. D. and van der Werf, J. H. 2012. The importance of information on relatives for the prediction of genomic breeding values and the implications for the makeup of reference data sets in livestock breeding schemes. *Genet. Sel. Evol.* 44: 10–1186.
de Bem Oliveira, I., Amadeu, R. R., Ferrão, L. F. V. and Muñoz, P. R. 2020. Optimizing whole-genomic prediction for autotetraploid blueberry breeding. *Heredity* 125: 437–448.
Dimitrijevic, A. and Horn, R. 2018. Sunflower hybrid breeding: from markers to genomic selection. *Frontiers in Plant Science* 8: 2238.
Falconer, D. S., Mackay, T. F. and Frankham, R. 1996. Introduction to quantitative genetics (4th edn). *Trends in Genetics* 12: 280.
Fristche-Neto, R., Akdemir, D. and Jannink, J.-L. 2018. Accuracy of genomic selection to predict maize single-crosses obtained through different mating designs. *Theoretical and Applied Genetics* 131: 1153–1162.
Fritsche-Neto, R., Galli, G., Alves, F., Sabadin, F., Lyra, D., Costa-Neto, G., Morais, P., Granato, I., Borges, K. L. and Crossa, J. 2021. Optimizing genomic-enabled prediction in small-scale low budged maize hybrid breeding programs: a roadmap review. *Frontiers of Plant Science.*
Goddard, M. 2009. Genomic selection: prediction of accuracy and maximisation of long term response. *Genetics* 136: 245–257.
Guo, T., Yu, X., Li, X., Zhang, H., Zhu, C., Flint-Garcia, S., McMullen, M. D., Holland, J. B., Szalma, S. J., Wisser, R. J. et al. 2019. Optimal designs for genomic selection in hybrid crops. *Molecular Plant* 12: 390–401.
Guo, Z., Tucker, D., Basten, C., Gandhi, H., Ersoz, E., Guo, B., Xu, Z., Wang, D. and Gay, G. 2014. The impact of population structure on genomic prediction in stratified populations. *Theoretical and Applied Genetics* 127: 749–762.
Habier, D., Fernando, R. and Dekkers, J. 2007. The impact of genetic relationship information on genome-assisted breeding values. *Genetics* 177: 2389–2397.

Habier, D., Tetens, J., Seefried, F.-R., Lichtner, P. and Thaller, G. 2010. The impact of genetic relationship information on genomic breeding values in german holstein cattle. *Genetics Selection Evolution* 42: 5.

Hayes, B., Bowman, P., Chamberlain, A. and Goddard, M. 2009. Invited review: Genomic selection in dairy cattle: Progress and challenges. *Journal of Dairy Science* 92: 433–443.

He, S., Schulthess, A. W., Mirdita, V., Zhao, Y., Korzun, V., Bothe, R., Ebmeyer, E., Reif, J. C. and Jiang, Y. 2016. Genomic selection in a commercial winter wheat population. *Theoretical and Applied Genetics* 129: 641–651.

Heslot, N. and Feoktistov, V. 2020. Optimization of selective phenotyping and population design for genomic prediction. *Journal of Agricultural, Biological and Environmental Statistics* 25: 579–600.

Isidro, J., Jannink, J.-L., Akdemir, D., Poland, J., Heslot, N. and Sorrells, M. E. 2015. Training set optimization under population structure in genomic selection. *Theoretical and Applied Genetics* 128: 145–158.

Isidro, J., Akdemir, D. and Burke, J. 2016. Genomic selection. pp. 1001–1023. *In*: William, A., Alain, B. and Maarten, V. G. (eds.). *The World Wheat Book: A History of Wheat Breeding*. Lavoisier, Paris.

Jannink, J.-L. 2010. Dynamics of long-term genomic selection. *Genetics Selection Evolution* 42: 35.

Kadam, D. C., Potts, S. M., Bohn, M. O., Lipka, A. E. and Lorenz, A. J. 2016. Genomic prediction of single crosses in the early stages of a maize hybrid breeding pipeline. *G3: Genes, Genomes, Genetics* 6: 3443–3453.

Laloë, D. 1993. Precision and information in linear models of genetic evaluation. *Genetics Selection Evolution* 25: 557–576.

Liu, X., Wang, H., Wang, H., Guo, Z., Xu, X., Liu, J., Wang, S., Li, W.-X., Zou, C., Prasanna, B. M. et al. 2018. Factors affecting genomic selection revealed by empirical evidence in maize. *The Crop Journal* 6: 341–352.

Longin, C. F. H., Mi, X. and Würschum, T. 2015. Genomic selection in wheat: optimum allocation of test resources and comparison of breeding strategies for line and hybrid breeding. *Theoretical and Applied Genetics* 128: 1297–1306.

Lorenz, A. J. and Smith, K. P. 2015. Adding genetically distant individuals to training populations reduces genomic prediction accuracy in barley. *Crop Science* 55: 2657–2667.

Mangin, B., Bonnafous, F., Blanchet, N., Boniface, M.-C., Bret-Mestries, E., Carrère, S., Cottret, L., Legrand, L., Marage, G., Pegot-Espagnet, P. et al. 2017. Genomic prediction of sunflower hybrids oil content. *Frontiers in Plant Science* 8: 1633.

Mangin, B., Rincent, R., Rabier, C.-E., Moreau, L. and Goudemand-Dugue, E. 2019. Training set optimization of genomic prediction by means of ethacc. *PloS One* 14: e0205629.

Marulanda, J. J., Mi, X., Melchinger, A. E., Xu, J.-L., Würschum, T. and Longin, C. F. H. 2016. Optimum breeding strategies using genomic selection for hybrid breeding in wheat, maize, rye, barley, rice and triticale. *Theoretical and Applied Genetics* 129: 1901–1913.

McClellan, J., Susser, E. and King, M. 2007. Schizophrenia: a common disease caused by multiple rare alleles. *The British Journal of Psychiatry* 190: 194–199.

Meuwissen, T., Hayes, B. and Goddard, M. 2001. Prediction of total genetic value using genome-wide dense marker maps. *Genetics* 157: 1819–1829.

Neyhart, J. L., Tiede, T., Lorenz, A. J. and Smith, K. P. 2017. Evaluating methods of updating training data in long-term genomewide selection. *G3: Genes, Genomes, Genetics* 7: 1499–1510.

Norman, A., Taylor, J., Edwards, J. and Kuchel, H. 2018. Optimising genomic selection in wheat: Effect of marker density, population size and population structure on prediction accuracy. *G3: Genes, Genomes, Genetics* 8: 2889–2899.

Olatoye, M. O., Clark, L. V., Labonte, N. R., Dong, H., Dwiyanti, M. S., Anzoua, K. G., Brummer, J. E., Ghimire, B. K., Dzyubenko, E., Dzyubenko, N. et al. 2020. Training population optimization for genomic selection in miscanthus. *G3: Genes, Genomes, Genetics* 10: 2465–2476.

Ou, J.-H. and Liao, C.-T. 2019. Training set determination for genomic selection. *Theoretical and Applied Genetics* 132: 2781–2792.

Pszczola, M., Strabel, T., Mulder, H. and Calus, M. 2012. Reliability of direct genomic values for animals with different relationships within and to the reference population. *Journal of Dairy Science* 95: 389–400.

Reif, J. C., Zhao, Y., Würschum, T., Gowda, M. and Hahn, V. 2013. Genomic prediction of sunflower hybrid performance. *Plant Breeding* 132: 107–114.

Rincent, R., Laloë, D., Nicolas, S., Altmann, T., Brunel, D., Revilla, P., Rodriguez, V. M., Moreno-Gonzalez, J., Melchinger, A., Bauer, E. et al. 2012. Maximizing the reliability of genomic selection by optimizing the calibration set of reference individuals: comparison of methods in two diverse groups of maize inbreds (*Zea mays* L.). *Genetics* 192: 715–728.

Rincent, R., Charcosset, A. and Moreau, L. 2017. Predicting genomic selection efficiency to optimize calibration set and to assess prediction accuracy in highly structured populations. *Theoretical and Applied Genetics* 130: 2231–2247.

Roth, M., Muranty, H., Di Guardo, M., Guerra, W., Patocchi, A. and Costa, F. 2020. Genomic prediction of fruit texture and training population optimization towards the application of genomic selection in apple. *Horticulture Research* 7: 1–14.

Sarinelli, J. M., Murphy, J. P., Tyagi, P., Holland, J. B., Johnson, J. W., Mergoum, M., Mason, R. E., Babar, A., Harrison, S., Sutton, R. et al. 2019. Training population selection and use of fixed effects to optimize genomic predictions in a historical USA winter wheat panel. *Theoretical and Applied Genetics* 132: 1247–1261.

Schrag, T., Frish, M., Dhillon, B. and Melchinger, A. 2009. Marker-based prediction of hybrid performance in maize single-crosses involving doubled haploids. *Maydica* 54: 353.

Schulthess, A. W., Zhao, Y. and Reif, J. C. 2017. Genomic selection in hybrid breeding. pp. 149–183. *In*: Varshney, Rajeev K., Roorkiwal, Manish and Sorrells, Mark E. (eds.). *Genomic Selection for Crop Improvement.* Springer.

Seye, A., Bauland, C., Charcosset, A. and Moreau, L. 2020. Revisiting hybrid breeding designs using genomic predictions: simulations highlight the superiority of incomplete factorials between segregating families over topcross designs. *Theoretical and Applied Genetics* 133: 1995–2010.

Technow, F., Schrag, T. A., Schipprack, W., Bauer, E., Simianer, H. and Melchinger, A. E. 2014. Genome properties and prospects of genomic prediction of hybrid performance in a breeding program of maize. *Genetics* 197: 1343–1355.

Wang, Y., Mette, M. F., Miedaner, T., Gottwald, M., Wilde, P., Reif, J. C. and Zhao, Y. 2014. The accuracy of prediction of genomic selection in elite hybrid rye populations surpasses the accuracy of marker-assisted selection and is equally augmented by multiple field evaluation locations and test years. *BMC Genomics* 15: 1–12.

Wientjes, Y. C., Veerkamp, R. F. and Calus, M. P. 2013. The effect of linkage disequilibrium and family relationships on the reliability of genomic prediction. *Genetics* 193: 621–631.

Windhausen, V. S., Atlin, G. N., Hickey, J. M., Crossa, J., Jannink, J.-L., Sorrells, M. E., Raman, B., Cairns, J. E., Tarekegne, A., Semagn, K. et al. 2012. Effectiveness of genomic prediction of maize hybrid performance in different breeding populations and environments. *G3: Genes, Genomes, Genetics* 2: 1427–1436.

Xu, Y., Liu, X., Fu, J., Wang, H., Wang, J., Huang, C., Prasanna, B. M., Olsen, M. S., Wang, G. and Zhang, A. 2020. Enhancing genetic gain through genomic selection: from livestock to plants. *Plant Communications* 1: 100005.

Yu, X., Leiboff, S., Li, X., Guo, T., Ronning, N., Zhang, X., Muehlbauer, G. J., Timmermans, M. C., Schnable, P. S., Scanlon, M. J. et al. 2020. Genomic prediction of maize microphenotypes provides insights for optimizing selection and mining diversity. *Plant Biotechnology Journal* pp. 2456–2465.

Zhang, H., Yin, L., Wang, M., Yuan, X. and Liu, X. 2019. Factors affecting the accuracy of genomic selection for agricultural economic traits in maize, cattle, and pig populations. *Frontiers in Genetics* 10: 189.

Zhao, Y., Mette, M., Gowda, M., Longin, C. and Reif, J. 2014. Bridging the gap between marker-assisted and genomic selection of heading time and plant height in hybrid wheat. *Heredity* 112: 638–645.

Zhao, Y., Mette, M. F. and Reif, J. C. 2015. Genomic selection in hybrid breeding. *Plant Breeding* 134: 1–10.

Zhao, Y., Zeng, J., Fernando, R. and Reif, J. C. 2013. Genomic prediction of hybrid wheat performance. *Crop Science* 53: 802–810.

Zhong, S., Dekkers, J. C., Fernando, R. L. and Jannink, J.-L. 2009. Factors affecting accuracy from genomic selection in populations derived from multiple inbred lines: a barley case study. *Genetics* 182: 355–364.

3

Genomic Selection in Wheat: Progress, Opportunities and Challenges

Deepmala Sehgal, Suchismita Mondal, Juan Burgeño, Umesh Rosyara, Alison R Bentley* and *Susanne Dreisigacker*

ABSTRACT

The upsurge in sequencing and genotyping technologies has enabled the mainstreaming of genomic selection (GS) in plant breeding programs globally. It is one of the leading approaches in efforts to accelerate genetic gain. In wheat, various approaches and models have been tested for improving prediction accuracies for complex traits including optimizing the size of the training population, relationships between individuals, marker type and density and by use of pedigree information, environmental covariates and other parameters. The progress in high-throughput, detailed imaging and phenotyping systems has provided an opportunity to improve prediction accuracies using high quality secondary traits. This chapter will review the advances made in GS-based breeding strategies for wheat breeding with a special focus on CIMMYT's accomplishments in this area. The opportunities and challenges of effectively implementing GS routinely in a global wheat program are also discussed.

Introduction

With the advent of molecular markers in 1990s, it was envisioned that molecular breeding would progress at a rapid pace through the application

International Maize and Wheat Improvement Center (CIMMYT), Km. 45, Carretera Méx-Veracruz, El Batán, Texcoco, México, CP 56237.

* Corresponding author: d.sehgal@cgiar.org

of marker-assisted selection (MAS). For traits controlled by major genes or major quantitative trait loci (QTL), this became a reality and many publications document the products and varieties arising from the use of MAS in many crops (Shanti et al., 2010; Katula-Debreceni et al., 2010; Zhao et al., 2012). Shanti et al. (2010), for example, pyramided four bacterial blight resistance genes (*Xa*5, *Xa*4, *xa*13 and *Xa*21) into hybrid rice parental lines using simple sequence repeat (SSR) markers with the resulting pyramided lines showing high levels of disease resistance to multiple highly virulent isolates of *Xanthomonas oryzae* pv. *oryzae*. Similarly, Zhao et al. (2012) introgressed *qHSR1* (a major QTL for head smut resistance) using SSR markers to improve maize resistance to head smut. In wheat, marker-assisted backcrossing (MABC), forward breeding and F_2 enrichment have all been used to successfully introgress major genes (e.g., Dubcovsky, 2004; Bonnett et al., 2005; Kuchel et al., 2005; 2007). Interestingly, 80 MAS and 350 MABC projects have been reported in wheat (https://maswheat.ucdavis.edu/), transferring major rust and quality genes into 75 different recurrent parents (Dubcovsky, 2004; Kuchel et al., 2007).

However, for improvement of complex traits that are controlled by multiple small-effect loci displaying significant loci × loci (epistatsis) (Sehgal et al., 2017; 2019) and genotype × environment interactions, MAS alone is unlikely to be a viable option. This means that new selection strategies such as genomic selection (GS) may pave the way for achieving further genetic gains based on genomic information. GS is a modification of MAS where marker effects across the entire genome are estimated to calculate genomic estimated breeding values (GEBVs; Meuwissen et al., 2001) using a prediction model with selection of individuals based on GEBVs. One of the earliest proofs that real genetic gains are achievable through GS in wheat came from the work of Heffner et al. (2010; 2011a, b). They compared conventional phenotypic and MAS prediction accuracy with genomic prediction for grain quality and 13 other agronomic traits in two biparental populations (Heffner et al., 2011a) and multifamily data (Heffner et al., 2011b) from the Cornell University wheat breeding program. In biparental populations, the authors showed an average ratio of GS accuracy to phenotypic selection of 0.66 while a 28% increase in prediction accuracy was observed in the multifamily design using GS over MAS. These studies indicate that the GS approach can be an alternative to conventional phenotypic selection or MAS.

In the post-genomics era where sequencing costs are declining, deployment of genomics-assisted breeding approaches have become common in many breeding programs. For wheat, availability of dense markers was a major limitation in conducting in-depth genetic analyses as compared to other major crops such as rice and maize. This is largely due to large (16 Gb) genome size and its hexaploid nature. The availability of dense sets of single-nucleotide polymorphisms (SNPs) in the past decade (arising from a range of different genotyping platforms such as 90K Illumina iSelect, genotyping by sequencing (GBS), DArTseq, high-density Affymetrix Axiom® genotyping array) has brought a significant transformation in the marker tool kit available for wheat (Poland et al., 2012; Cavanagh et al., 2013; Allen et al., 2017). The resulting high-density genomic data have opened new possibilities in wheat to explore the genetic architecture of traits using genome-wide association study (GWAS) to link traits to markers, and to perform other genomic studies, for instance, the analysis of selective sweeps within or across species (Afzal et al., 2019; Liu et al., 2019). Concomitantly, the availability of dense marker data has allowed the testing and validation of various GS models for several traits using different marker types and densities and by including pedigrees, environmental covariates and high throughput phenotyping data. This chapter will review the recent advancements made in research in this area in wheat breeding programs globally with a special focus on CIMMYT's accomplishments in this area.

Optimizing training population design and size

The training population (TP) is a set of individuals that has been both genotyped and phenotyped in order to train the statistical genomic prediction model. Once the model is trained, it takes genotypic information from a population that lacks phenotypic data, often called the test or validation set, and produces GEBVs for selection. Design and size of TP is critical to achieving acceptable levels of accuracy and therefore their optimization has received much interest in both animal and plant breeding. The setup of an initial TP can be difficult and is often a large investment since breeders have to take into account multiple factors to decide the number of lines to be genotyped and phenotyped to establish a suitable training data set. Various types of training populations have been used in wheat, ranging from bi-parental and tri-parental populations (Ornella et al., 2012; Heffner et al., 2011a; Muleta et al., 2017), multi-lines (Cross et al., 2010; Saint Pierre et al., 2016; Hoffsetter et al., 2016; Juliana et al., 2017a, b; Dong et al., 2018; Ladejobi et al., 2019), full-sib and half-sib (Cericola et al., 2019) to multi-subpopulations or multi-families (Heffner et al., 2011b; Edwards et al., 2019) and even gene bank accessions (Daetwyler et al., 2014; Crossa et al., 2016; Velu et al., 2016; Phillip et al., 2018).

Crossa et al. (2016) were the first to use GS models on a large set of wheat landraces and accessions from a gene bank. They used 8416 Mexican and 2403 Iranian landrace accessions to predict days to heading (DTH) and days to maturity (DTM) by using two strategies. The first involved random cross-validation of the data in 20% training and 80% testing and the second involved two types of core sets, "diversity" and "prediction", including 10%

and 20%, respectively, of the total collections. Prediction accuracy of the 20% diversity core set was close to accuracies obtained for 20% training and 80% testing set (0.412 to 0.654 and 0.182 to 0.647 for Mexican landraces and Iranian landraces, respectively).

Investigations on TP size, using 100 to 1,000 individuals have shown a general trend that increasing TP size leads to an increase in prediction accuracy (Heffner et al., 2011a, b; Hickey et al., 2014; Bentley et al., 2014; Isidoro et al., 2015; Michel et al., 2017; Lozada et al., 2019; Sarinelli et al., 2019). Norman et al. (2018), however, used a large panel of 10,375 bread wheat lines to investigate the effect of TP sizes on genomic prediction accuracy. The authors divided their panel into TP sizes from 250 to 8,300 lines with differing relationships and found that the increase in prediction accuracy slowed beyond 2,000 lines. This study provided most relevant results for large scale breeding programs working with thousands of lines because the results provide evidence on which to base the selection of minimum TP size.

Many studies have focused on the impact of population structure (Isidro et al., 2014; Crossa et al., 2016; Norman et al., 2018; Sarinelli et al., 2019) and the size of TP on prediction accuracy. These investigations have established that without accounting for population structure there can be preferential selection of individuals within a single subpopulation, leading to a gradual loss of diversity from the breeding program. Isidro et al. (2014) evaluated five different sampling algorithms for selecting TP size in the presence of different levels of population structure in order to investigate the effects on prediction accuracy for grain yield, test weight, lodging, heading date and plant height in wheat. The five algorithms included stratified sampling, mean of the coefficient of determination (CDmean), mean of predictor error variance (PEVmean), stratified CDmean (StratCDmean) and random sampling. The dataset included a population of 1,127 soft winter wheat varieties and F_5-derived advanced breeding lines resulting from many different crosses. Principal components analysis (PCA) was used to determine population structure. Their results indicated that it is difficult to choose a single best criterion because of the interaction of the trait architecture with population structure. However, the stratified sampling method showed higher accuracies than other methods when the trait was highly affected by population structure (or there was a strong structure in the TP) and CDmean and StratCDmean showed better accuracies when population structure effects were mild. Crossa et al. (2016), on the other hand, found a decrease in prediction accuracy of 15–20% when population structure was accounted for in the prediction models.

Approaches for prediction of wheat traits

The upsurge in marker data density coming from different genotyping platforms presents the challenge not only of handling large datasets, but also in dealing with the problem of p >> n, i.e., number of markers being 100 or 1000 times higher than the number of entries. To overcome this issue, commonly called 'curse of dimensionality', many models and algorithms have been developed for GS in both plant and animal research which deal with markers and/or additional parameters differently. The most common models include ridge regression–best linear unbiased prediction (RR-BLUP) and Bayesian-based methods such as BayesA, BayesB, BayesCπ, and BayesLASSO. An important difference between RR-BLUP and the Bayesian methods is the prior distribution for the variance of marker effects. RR-BLUP assigns equal variance to all markers and thus shrinks each marker effect equally toward zero, while Bayesian-based methods allow unequal variances for markers. In wheat, various models and their modifications have been investigated for improving prediction accuracies for traits with different architectures. Below, we review some of these results and detail the most relevant results.

Modeling marker number, density, and type

Studies conducted on marker number and densities have revealed that by using approximately 1000 genome-wide, well-spaced markers, maximum prediction accuracy can be reached in closely related wheat breeding lines (Heffner et al., 2011b; Jiang et al., 2015; Cercola et al., 2017; Muleta et al., 2017). A greater marker density is required only when the objective is to predict highly unrelated lines, i.e., that are distantly or not at all related (Norman et al., 2018). Combs and Bernardo (2013) showed that once the genome is sufficiently saturated with a marker for every 4.5 cM in wheat, the gain in prediction accuracy plateaued. Norman et al. (2018) investigated the effect of marker density and its interaction with population structure using marker subsets containing between 100 and 17,181 markers. Their results suggested that high marker density is very important when using a diverse training set to predict between poorly related material.

Many investigations have been conducted using only subsets of markers significantly associated with quantitative trait loci (QTL) in animals. These studies have obtained a consistency in results indicating that small subsets of markers (based on allele effects) can give similar accuracy to that obtained using all markers (Moser et al., 2010; Abdollahi-Arpanahi et al., 2014). However, it must be noted that in these cases whole genome marker sets in the range from 0.1 to 0.3 million SNPs were used and sub-setting SNPs linked to QTL would still be large enough to cover the genome and/or to provide sufficient marker densities. Similar investigations are limited in wheat. For instance, Hoffsetter et al. (2016) showed that systematically reducing the number of markers in the model to include only those markers significantly associated with the trait at $P < 0.05$ (i.e., 4.7–8.7% of all the markers) increase the prediction accuracy for most traits from 39 to 83%. The authors suggested based on their results that including markers in a GS model that are not significantly associated with a QTL can lower the accuracy because the effects associated with these regions are poorly estimated and hence these markers add noise to the data set. Lozada et al. (2019), however, obtained contradictory results to that of Hoffsetter et al. (2016). They tested the effects of marker subsets on GS accuracy and subsets of markers were selected based on different levels of significance, i.e., *P* values in genome-wide association analysis for yield components; subset $SS_{0.15}$ ($P < 0.15$), $SS_{0.10}$ ($P < 0.10$), and $SS_{0.05}$ ($P < 0.05$). The authors observed a 14–39% decrease in accuracy using the marker subsets for predictions as compared to whole genome markers for yield components. Kristensen et al. (2018) conducted GWAS to identify SNPs associated with five important wheat quality traits (grain protein content, Zeleny sedimentation, test weight, thousand-kernel weight, and falling number) and SNPs with large effects were identified for Zeleny sedimentation but not for the other investigated traits. For all traits, three and ten best SNPs as marker subsets were chosen to be used in GS models and the predictions improved when using the ten best SNPs instead of the three best. However, predictions were highest for all traits when all 10,802 SNPs were used in the model as compared to three or ten best SNPs. Overall, these studies suggest that using a marker subset can be beneficial in prediction if it effectively covers the genome and/or where marker effects are large. However, for complex traits in wheat, further investigations are needed to assess these benefits.

Poland et al. (2012) compared prediction accuracy using genotyping-by-sequencing (GBS) markers with the array-based Diversity Array Technology (DArT) platform. They also compared four different imputation methods to assess their impacts on prediction accuracy finding that whilst the imputation methods did not result in any significant change in accuracy, GBS markers produced significantly more accurate GEBV values than DArT markers. Even with a lower set of GBS markers (1,729 markers), which was comparable to DArT (1,827 markers), the GBS gave gains of approximately 0.15 compared to DArT markers. Recently Ladejobi et al. (2019) reached as similar conclusion in the comparison of GBS and array-based markers.

Modelling genotype × environment (G × E) interactions and use of environmental co-variates and multivariate models

The interaction between genotype and environment (G × E) has a strong impact on the yield of major crop plants and hence incorporation of G × E interactions in GS models has been an important area of investigation. In several crop species (Ly et al., 2013; Wang et al., 2016), fitting G × E interaction effects in prediction models has shown to be more effective in improving trait predictions than univariate or single environment prediction models (Burgeno et al., 2012; Lopez-Cruz et al., 2015; Jarquin et al., 2014; Heslot et al., 2014). Burgueño et al. (2012) evaluated the impact of modeling G × E in wheat using CIMMYT historical lines assessed in several mega-environments. Their results indicated that compared to single-environment mixed models, the modelling of G × E with a factor analytic structure in a genomic best linear unbiased prediction (GBLUP) model increased prediction accuracy by 21.8%. Incorporation of G × E effects, especially multi-location, multi-year, and multi-treatment data has been shown to provide substantial improvements in prediction accuracies in wheat (Crossa et al., 2016; Pérez-Rodríguez et al., 2017; Heslot et al., 2014; Lopez-Cruz et al., 2015; Jarquín et al., 2017). Heslot et al. (2014) used environmental co-variables and marker data in their G × E model and showed an improvement in prediction by 11.1% on average. Jarquín et al. (2014) developed a reaction norm model, where the main and interaction effects of markers and environmental covariates are introduced using high-dimensional random variance-covariance structures of markers and environmental covariates. Their model was tested using data from 139 wheat lines genotyped with 2,395 SNPs and evaluated for grain yield over 8 years and various locations within northern France. A total of 68 environmental covariates, defined based on five phases of the phenology of the crop, were used in the analysis. The prediction accuracy of the model was 17–34% higher than that of models based on main effects only (Jarquin et al., 2014).

Lopez-Cruz et al. (2015) modeled G × E using a marker × environment interaction (M × E) GS model. The authors analyzed three CIMMYT wheat data sets which included advanced lines from three yield trials. These lines were evaluated under three irrigation regimes (2I = two irrigations to simulate moderate drought stress, 5I = five irrigations simulating an optimally irrigated environment, and drip = simulating high drought stress). The authors compared the M × E model with a standard across-environment GS model that assumes that effects are constant across environments (i.e., ignoring G × E). The M × E model decomposed marker effects into two components; stable across environments (main effects) and environment-specific (interactions). The prediction accuracy of the M × E model was 5–29% greater than the standard across-environment GS model.

Huang et al. (2016) suggested the use of a stability index, as a complement to modelling G × E interactions in prediction models for the prediction of complex traits like yield. The authors used Additive Main-effect and Multiplicative Interaction (AMMI) and Eberhart and Russell Regression (ERR) models to estimate trait stability for multiple traits including yield and yield components, phenology and quality in a soft winter wheat population of 273 lines. Although inconclusive results were obtained, the accuracy of trait stability was only greater than the trait value itself for yield (0.44 using AMMI versus 0.33) and heading date (0.65 using ERR versus 0.56).

Ibba et al. (2020) and Gill et al. (2021) recently explored multivariate models for their performance in advanced breeding trials. Ibba et al. (2021) evaluated two multi-trait models (Bayesian multi-trait multi-environment [BMTME] and multi-trait ridge regression [MTR]) for predicting 13 wheat quality traits using a large set of wheat lines. Using these models, the authors predicted the performance of the lines in second year yield trials (used as testing set) using the quality traits data obtained in the first year of trialing (used as training set). The BMTME model was found to result in the best predictions for the majority of the traits as compared to the MTR model. The authors concluded that the BMTME model, which utilizes both correlation among traits and environments, should be used for multi-trait prediction analyses. Gill et al. (2021) used multi-trait multi-environment (MTME) GP models to predict for GY, grain protein content (GPC), test weight (TW), plant height (PH), days to heading (DH) using 314 advanced and elite winter wheat breeding lines and compared its performance with a single trait model. The authors reported explicitly the superior performance of the MTME model for all traits as compared to single trait model, with a significant improvement of up to 19, 71, 17, 48, and 51% for GY, GPC, TW, PH and DH, respectively.

Models integrating pedigree alone, pedigree and markers together and/or along with epistatic effects

Use of pedigree alone or in combination with markers in GS models has been investigated in detail by CIMMYT scientists in order to take advantage of the high quality and regularly updated pedigree data maintained by the breeding programs. Crossa et al. (2010) investigated two different models [GBLUP with the linear genomic matrix G and a nonparametric model Reproducing Kernel Hilbert Spaces (RKHS) regression with a nonlinear genomic matrix, the Gaussian kernel (GK)] using pedigree and genomic information on a collection of 599 wheat lines. The RKHS model, including pedigree and marker information, gave the highest prediction accuracy for grain yield, ranging from 15% to 36%, as compared to the pedigree model alone. The RKHS model incorporating markers and pedigree data was also found to be superior when compared with standard least-squares multiple regression methods for predicting resistance to leaf stem, and stripe rust, septoria, tan spot, and *Stagonospora nodorum* blotch in wheat (Juliana et al., 2017a, b).

Crossa et al. (2014) combined pedigree-derived additive and epistatic additive × additive interactions in GS models for predicting grain yield (GY). The models including epistatic effects increased the prediction accuracy in three of the four environments investigated. Pérez-Rodríguez et al. (2017)

utilized genotypic (GBS) and pedigree data for 58,798 CIMMYT wheat lines and used single-step genomic and pedigree models to predict the GY performance of lines in several sites in South Asia (India, Pakistan, and Bangladesh) using the reaction norm model developed by Jarquín et al. (2014) for incorporating G × E. Their results indicated that the prediction accuracy achieved by models including only pedigree, markers or both is higher (0.25–0.38) than prediction accuracy of models that use only phenotypic prediction (0.20) or do not include the G × E term.

Machine learning models

With the increases in volumes of data, machine learning (ML) and deep learning methods have become popular and hold potential for conducting multiple complex analyses. In wheat, the support vector regression (SVR) method, which is considered as the state-of-the-art, nonparametric, machine-learning algorithm, was compared with the performance of two linear models, ridge regression (rrBLUP) and Bayesian Lasso (BL), for predicting stem rust (SR) and yellow rust (YR) resistance in five wheat populations (Ornella et al., 2012). Overall, the BL and rrBLUP had similar prediction performance and these two linear models outperformed the two SVR models in most data sets. The superior prediction ability of the BL and rrBLUP over SVR based models was suggested to be due to the large additive effects of the associated loci for the two traits. However, for complex traits such as grain yield, Long et al. (2011b) found SVR models, considering both additive and epistatic effects, to be superior giving a 17.5% increase in correlation between predicted and observed values over linear model BL. González-Camacho et al. (2018) investigated two support vector classification (SVC) methods [Gaussian (SVC-g) and linear (SVC-l) kernels] with RKHS, BL and rrBLUP methods to predict for rust resistance and grain yield in 16 wheat data sets. SVC-l gave the highest prediction accuracy in almost all YR datasets and in about four SR datasets. RKHS gave the best values for grain yield in most data sets. More research is required on the use of ML models for predicting complex traits with different architectures to ascertain the advantages, if any, offered by new algorithms implemented in these models.

Models integrating major genes or genome-wide association results as fixed effects

In animal breeding research, it has been demonstrated that the integration of prior QTL or gene-based information in GS models as fixed effects can result in increased prediction accuracy, especially for traits with complex genetic architecture (Boichard et al., 2012; de los Campos et al., 2013; Su et al., 2014; Zhang et al., 2014; Veroneze et al., 2016; Lopes et al., 2017). In crops, this approach is yet to be investigated in detail. In wheat, Rutkoski et al. (2014) assessed the role of major rust genes and adult plant resistance (APR) loci as fixed effects in improving predictions accuracy of GS models for stem rust resistance. They identified six GBS markers associated with APR in a genome wide association study (GWAS). A KASP marker tightly linked to *Sr2* gene along with six APR loci were used as fixed effects in a G-BLUP model. Although gain in prediction accuracy was not dramatically increased, from $r = 0.57$ in GBLUP to $r > 0.6$ using G-BLUP with *Sr2*-linked markers as fixed effects, the results showed some increase in predictive performance. Later, Juliana et al. (2017a, b) also used this approach in wheat but did not find any significant improvement of the fixed effects model over RKHS model.

Sarinelli et al. (2019) recently tested the use of the fixed effects approach while optimizing GS models in a historical winter wheat panel. They included markers associated with major genes as fixed effects in prediction models for heading date, plant height, and resistance to powdery mildew. Improvement in predictive ability of the GS model was observed when major genes were added as fixed effects and the highest accuracies were observed when multiple gene combinations were incorporated into the model. Predictive ability of models having *Rht-D1* and *Rht-B1* ranged from 9 to 17% higher compared to the models without markers as fixed effects. Similarly, an average of 10% increase in model predictive ability was observed with a TP size of 50 when markers in the *Vrn-A1*, *Vrn-B1,* and *Ppd-D1* genes were included as fixed effects in the model. However, improvement in predictive ability was only 3% when the TP size was 350. Odilbekov et al. (2019) used the same approach of fixed effects in predicting for resistance to septoria blotch in Nordic winter wheat at seedling stage. The prediction accuracy increased from 0.47 to 0.62 when ten significant markers were used as fixed effects in prediction models.

Sehgal et al. (2020) used haplotyped-based GWAS loci as fixed effects in genomic prediction models to test their effects on prediction accuracies for complex traits. The authors specifically explored whether integrating consistent and robust associations identified in GWAS as fixed effects in prediction models improves prediction accuracy for grain yield and yield stability CIMMYT spring wheat. It was reported that the model accounting for the haplotype-based GWAS loci as fixed effects led to up to 9–10% increase in prediction accuracy for grain yield.

Role of integrating high throughput imaging technologies

With genotyping costs becoming cheaper and more affordable for large-scale application in breeding, substantial efforts have shifted to the development of high-throughput phenotyping (HTP) platforms. Across crops, including wheat, these platforms present the opportunity to generate large-scale and in-depth phenotyping data (Zhang and Zhang, 2018; Walter et al., 2019). A range of HTP platforms are being used to measure different traits in wheat, ranging from plant height (Holman et al., 2016), disease resistance (Devadas et al., 2015), growth rate (Holman et al., 2016) to different vegetation indices (Haghighattalab et al., 2016). Moreover, different models have been investigated to extract the large volume and dimensionality of data collected from HTP platforms (Rutkoski et al., 2016; Sun et al., 2017). Utilization of HTP data in GS models is gaining a lot of attention in wheat and significant advancements have been made in this area of research (Rutkoski et al., 2016; Sun et al., 2019; Juliana et al., 2019). Rutkoski et al. (2016) investigated the role of canopy temperature and green and red normalized difference vegetation index as secondary traits in GS models for improving prediction accuracy for grain yield using 557 CIMMYT lines. These lines were evaluated in five contrasting environments and predictions were evaluated for each environment. The authors modeled the two secondary traits on training and test sets, and grain yield on the training set multivariate and observed 67% improvement in prediction accuracy without correcting for days to heading (DTH). When the authors corrected for DTH, the models still showed an improvement of 37%. Similarly, Crain et al. (2018) utilized HTP to evaluate 1170 advanced CIMMYT lines in drought and heat stress environments. They tested different GS models using 2,254 GBS markers and over 1.1 million HTP phenotypic datapoints. The secondary traits were modelled as a response in multivariate models or as a covariate in univariate models. An increase in prediction accuracy from 7 to 33% above the standard univariate model was obtained. Juliana et al. (2019) performed multivariate prediction of GY using the green normalized difference vegetation index from HTP and achieved average accuracies of 0.56 and 0.62 for GY in the drought and heat stress environments, respectively.

Conclusions

In summary, the advancements in genomics technologies leading to a plethora of high-density markers has significantly increased the use and potential of GS in crops, including wheat. The studies on GS highlighted in this chapter have led to the following important conclusions: (1) larger TPs usually result in a better prediction accuracies, (2) a closer relationship between training and test populations leads to higher accuracy, (3) incorporating G × E interactions into GS models improves prediction accuracy of complex traits in previously untested environments, (4) incorporating pedigree information along with markers in GS models gives higher accuracy than markers alone and (5) Integrating HTP data in multivariate models increases prediction accuracy.

Hence, a wealth of information is available on the prediction accuracies of different GS models for multiple traits with different architectures. While most of the research has shown moderate to high prediction accuracies in wheat, more work is required to incorporate it routinely in wheat breeding programs globally. At CIMMYT, a large GBS data set combined with agronomic, grain yield and high-throughput phenotyping data has been generated on 80,000 spring wheat breeding lines and has been used in GS models to predict for multiple traits with different architectures. In addition, rapid-cycle GS is being tested to exploit the full advantage of genomic prediction for increasing genetic gains per unit of time.

References

Abdollahi-Arpanahi, R., Nejati-Javaremi, A., Pakdel, A., Moradi-Shahrbabak, M., Morota, G., Valente, B. D., Kranis, A., Rosa, G. J. M. and Gianola, D. 2014. Effect of allele frequencies, effect sizes and number of markers on prediction of quantitative traits in chickens. *J. Anim. Breed. Genet.* 131: 123–133.

Afzal, F., Li, H., Gul Kazi, A., Subhani, A., Ahmad, A., Mujeeb-Kazi, A., Ogbonnaya, F., Trethowan, R., Xia, X., He, Z. and Rasheed, A. 2019. Genome-wide analyses reveal footprints of divergent selection and drought adaptive traits in synthetic-derived wheats. *G3 Genes Gen. Genet.* 9: 1957–1973.

Allen, A. M., Winfield, M. O., Burridge, A. J., Downie, R. C., Benbow, H. R., Barker, G. L., Wilkinson, P. A., Coghill, J., Waterfall, C., Davassi, A., Scopes, G., Pirani, A., Webster, T., Brew, F., Bloor, C., Griffiths, S., Bentley, A. R., Alda, M., Jack, P., Phillips, A. L. and Edwards, K. J. 2017. Characterization of a Wheat Breeders' Array suitable for high-throughput SNP genotyping of global accessions of hexaploid bread wheat (*Triticum aestivum*). *Plant Biotech. J.* 15(3): 390–401.

Bentley, A. R., Scutari, M., Gosman, N., Faure, S., Bedford, F., Howell, P., Cockram, J., Rose, G. A., Barber, T., Irigoyen, J., Horsnell, R., Pumfrey, C., Winnie, E., Schacht, J., Beauchene, K., Praud, S., Greenland, A., Balding, D. and Mackay, I. J. 2014. Applying association mapping and genomic selection to the dissection of key traits in elite European wheat. *Theor. Appl. Genet.* 127: 2619–2633.

Boichard, D., Guillaume, F., Baur, A., Croiseau, P., Rossignol, M. N., Boscher, M. Y., Druet, T., Genestout, L., Eggen, A., Journaux, L., Ducrocq, V. and Fritz, S. 2012. Genomic selection in French dairy cattle. *Anim. Prod. Sci.* 52: 115–20.

Bonnett, D. G., Rebetzke, G. J. and Spielmeyer, W. 2005. Strategies for efficient implementation of molecular markers in wheat breeding. *Mol. Breed.* 15: 75–85.

Burgueño, J., de los Campos, G., Weigel, K. and Crossa, J. 2012. Genomic prediction of breeding values when modeling genotype × environment interaction using pedigree and dense molecular markers. *Crop Sci.* 52: 707–719.

Cavanagh, C. R., Chao, S., Wang, S., Huang, B. E., Stephen, S., Kiani, S., Forrest, K., Saintenac, C., Brown-Guedira, G. L., Akhunova, A., See, D., Bai, G., Pumphrey, M., Tomar, L., Wong, D., Kong, S., Reynolds, M., Lopez da Silva, M., Bockelman, H., Talbert, L., Anderson, J. A., Dreisigacker, S., Baenziger, S., Carter, A., Korzun, V., Morrell, P. L., Dubcovsky, J., Morell, M. K., Sorrells, M. E., Hayden, M. J. and Akhunov, E. 2013. Genome-wide comparative diversity uncovers multiple targets of selection for improvement in hexaploid wheat landraces and cultivars. *Proc. Natl. Acad. Sci. U.S.A.* 110: 8057–8062.

Cericola, F., Jahoor, A., Orabi, J., Andersen, J. R., Janss, L. L. and Jensen, J. 2017. Optimizing training population size and genotyping strategy for genomic prediction using association study results and pedigree information. A case of study in advanced wheat breeding lines. *PLoS ONE* 12: e0169606.

Combs, E. and Bernardo, R. 2013. Accuracy of genome wide selection for different traits with constant population size, heritability and number of markers. *Plant Gen.* 6: 1–7.

Crossa, J., Jarquín, D., Franco, J., Pérez-Rodríguez, P., Burgueño, J., Saint-Pierre, C., Vikram, P., Sansaloni, C., Petroli, C., Akdemir, D., Sneller, C., Reynolds, M., Tattaris, M., Payne, T., Guzman, C., Peña, R. J., Wenzl, P. and Singh, S. 2016. Genomic prediction of gene bank wheat landraces. *Genes Genomes Genet.* 6(7): 1819–34.

Daetwyler, H. D., Bansal, U. K., Bariana, H. S., Hayden, M. J. and Hayes, B. J. 2014. Genomic prediction for rust resistance in diverse wheat landraces. *Theor. Appl. Genet.* 127: 1795–1803. doi: 10.1007/s00122-014-2341-8.

de los Campos, G., Vazquez, A. I., Fernando, R., Klimentidis, Y. C. and Sorensen, D. 2013. Prediction of complex human traits using the genomic best linear unbiased predictor. *PLOS Genet.* 9: e1003608.

Devadas, R., Lamb, D. W., Backhouse, D. and Simpfendorfer, S. 2015. Sequential application of hyperspectral indices for delineation of stripe rust infection and nitrogen deficiency in wheat. *Precision Agric.* 16: 477–491.

Dong, H., Wang, R., Yuan, Y., Anderson, J., Pumphrey, M., Zhang, Z. and Chen, J. 2018. Evaluation of the potential for genomic selection to improve spring wheat resistance to fusarium head blight in the pacific northwest. *Front. Plant Sci.* 9: 911.

Dubcovsky, J. 2004. Marker assisted selection in public breeding programs: the wheat experience. *Crop Sci.* 44: 1895–1898.

Gill, H. S., Halder, J., Zhang, J., Brar, N. K., Rai, T. S., Hall, C., Bernardo, A., St Amand, P., Bai, G., Olson, E., Ali, S., Turnispeed, B. and Sehgal, S. K. 2021. Multi-trait multi-environment genomic prediction of agronomic traits in advanced breeding lines of winter wheat. *Front. Plant Sci.* 12: 709545.

Gonzalez-Camacho, J., Ornella, L., Pérez-Rodríguez, P., Gianola, D., Dreisigacker, S. and Crossa, J. 2018. Applications of machine learning methods to genomic selection in breeding wheat for rust resistance. *The Plant Genome* 11.

Gonzalez-Sanchez, A., Frausto-Solis, J. and Ojeda-Bustamante, W. 2014. Predictive ability of machine learning methods for massive crop yield prediction. *Spanish J. Agric. Res.* 12: 313–328.

Haghighattalab, A., González Pérez, L., Mondal, S., Singh, D., Schinstock, D., Rutkoski, J., Oritiz-Monasterio, I., Singh, R., Goodin, D. and Poland, J. 2016. Application of unmanned aerial systems for high throughput phenotyping of large wheat breeding nurseries. *Plant Methods* 12(1): 35.

Heffner, E. L., Lorenz, A. J., Jannink, J. L. and Sorrells, M. E. 2010. Plant breeding with genomic selection: Gain per unit time and cost. *Crop Sci.* 50: 1681–1690.

Heffner, E., Jannink, J. L. and Sorrells, M. 2011a. Genomic selection accuracy using multifamily prediction models in a wheat breeding program. *The Plant Gen.* 4: 65–75.

Heffner, E., Jannink, J. L., Iwata, H., Souza, E. and Sorrells, M. 2011b. Genomic selection accuracy for grain quality traits in biparental wheat populations. *Crop Sci.* 51: 2597–2606.

Heslot, N., Akdemir, D., Sorrells, M. E. and Jannink, J. L. 2013. Integrating environmental covariates and crop modeling into the genomic selection framework to predict genotype by environment interactions. *Theor. Appl. Genet.* 127: 463–480. doi: 10.1007/s00122-013-2231-5.

Hickey, J. M., Dreisigacker, S., Crossa, J., Hearne, S., Babu, R., Prassana, B. M., Grondona, M., Zambelli, A., Windhaussen, V., Mathews, K. and Gorjanc, G. 2014. Evaluation of genomic selection training population designs and genotyping strategies in plant breeding programs using simulation. *Crop Sci.* 54: 1476–1488.

Hoffsetter, A., Cabrera, A., Huang, M. and Sneller, C. 2016. Optimizing training population data and validation of genomic selection for economic traits in soft winter wheat. *Genes Genet. Genom.* 6(9): 2919–2928.

Holman, F. H., Riche, A. B., Michalski, A., Castle, M., Wooster, M. J. and Hawkesford, M. J. 2016. High throughput field phenotyping of wheat plant height and growth rate in field plot trials using UAV based remote sensing. *Remote Sens.* 8(12): 1031.

Huang, M., Cabrera, A., Hoffstetter, A., Griffey, C., Van Sanford, D., Costa, J., McKendry, A., Chao, S. and Sneller, C. 2016. Genomic selection for wheat traits and trait stability. *Theor. Appl. Genet.* 129(9): 1697–710.

Ibba, M. A., Crossa, J., Montesinos-López, O. A., Montesinos-López, A., Juliana, P., Guzman, C., Delorean, E., Dreisigacker, S. and Poland, J. 2020. Genome-based prediction of multiple wheat quality traits in multiple years. *The Plant Gen.* 13: e20034.

Isidro, J., Jannink, J. L., Akdemir, D., Poland, J., Heslot, N. and Sorrells, M. E. 2015. Training set optimization under population structure in genomic selection. *Theor. Appl. Genet.* 128(1): 145–158.

Jarquín, D., Crossa, J., Lacaze, X., Du Cheyron, P., Daucourt, J., Lorgeou, J., Piraux, F., Guerreiro, L., Perez, P., Calus, M., Burgueño, J. and de los Campos, G. 2014. A reaction norm model for genomic selection using high-dimensional genomic and environmental data. *Theor. Appl. Genet.* 127: 595–607.

Jarquín, D., Lemes da Silva, R., Gaynor, C., Poland, J., Fritz, A., Howard, R., Battenfield, S. and Crossa, J. 2017. Increasing genomic-enabled prediction accuracy by modeling genotype × environment interactions in Kansas wheat. *Plant Gen.* 10.

Jiang, Y., Zhao, Y., Rodemann, B., Plieske, J., Kollers, S., Korzun, V., Ebmeyer, E., Argillier, O., Hinze, M. and Ling, J. 2015. Potential and limits to unravel the genetic architecture and predict the variation of Fusarium Head blight resistance in European winter wheat (*Triticum aestivum* L.). *Heredity* 114: 318.

Juliana, P., Singh, R. P., Singh, P. K., Crossa, J., Huerta-Espino, J., Lan, C., Bhavani, S., Rutkoski, J. E., Poland, J. A., Bergstrom, G. C. and Sorrells, M. E. 2017. Genomic and pedigree-based prediction for leaf, stem, and stripe rust resistance in wheat. *Theor. Appl. Genet.* 130: 1415–1430.

Juliana, P., Singh, R. P., Singh, P. K., Crossa, J., Rutkoski, J. E., Poland, J. A., Bergstrom, G. C. and Sorrells, M. E. 2017. Comparison of models and whole-genome profiling approaches for genomic-enabled prediction of *Septoria tritici* blotch, *Stagonospora nodorum* blotch, and tan spot resistance in wheat. *The Plant Gen.* 10: 1–16.

Juliana, P., Poland, J., Huerta-Espino, J., Shrestha, S., Crossa, J., Crespo-Herrera, L., Toledo, F. H., Govindan, V., Mondal, S., Kumar, U., Bhavani, S., Singh, P. K., Randhawa, M. S., He, X., Guzman, C., Dreisigacker, S., Rouse, M. N., Jin, Y., Perez-Rodriguez, P., Montesinos-Lopez, O. A., Singh, D., Rehman, M. M., Marza, F. and Singh, R. P. 2019. Improving grain yield, stress resilience and quality of bread wheat using large-scale genomics. *Nat. Genet.* 51: 1530–1539.

Katula-Debreceni, D., Lencsés, A. K., Szőke, A., Veres, A., Hoffmann, S., Kozma, P., Kovács, L. G., Heszky, L. and Kiss, E. 2010. Marker-assisted selection for two dominant powdery mildew resistance genes introgressed into a hybrid grape population. *Sci. Horti.* 126: 448–453.

Kuchel, H., Guoyou, Y., Fox, R. and Jefferies, S. 2005. Genetic and economic analysis of a targeted marker-assisted wheat breeding strategy. *Mol. Breed.* 16: 67–78.

Kuchel, H., Fox, R., Reinheimer, J., Mosionek, L., Willey, N., Bariana. H. and Jefferies, S. 2007. The successful application of a marker-assisted wheat breeding strategy. *Mol. Breed.* 20: 295–308.

Ladejobi, O., Mackay, I. J., Poland, J., Praud, S., Hibberd, J. M. and Bentley, A. R. 2019. Reference Genome anchoring of high-density markers for association mapping and genomic prediction in European winter wheat. *Front Plant Sci.* 10: 1278.

Liu, J., Rasheed, A., He, Z., Imtiaz, M., Arif, A., Mahmood, T., Ghafoor, A., Siddiqui, S. U., Ilyas, M. K., Wen, W., Gao, F., Xie, C. and Xia, X. 2019. Genome-wide variation patterns between landraces and cultivars uncover divergent selection during modern wheat breeding. *Theor. Appl. Genet.* 132: 2509–2523.

Long, N., Gianola, D., Rosa, G. J. and Weigel, K. A. 2011. Application of support vector regression to genome-assisted prediction of quantitative traits. *Theor. Appl. Genet.* 123: 1065–1074.

Lopes, M. S., Bovenhuis, H., Van Son, M. N., Grind Flek, E. H., Knol, E. F. and Bastiaansen, J. W. M. 2017. Using markers with large effect in genetic and genomic predictions. *J. Anim. Sci.* 95: 59–71.

Lopez-Cruz, M., Crossa, J., Bonnett, D., Dreisigacker, S., Poland, J., Jannink, J. L., Singh, R. P., Autrique, E. and de los Campos, G. 2015. Increased prediction accuracy in wheat breeding trials using a marker × environment interaction genomic selection model. *G3 552 Genes Gen. Genet.* 5: 569.

Lozada D. N. and Carter, A. H. 2019. Accuracy of single and multi-trait genomic prediction models for grain yield in US Pacific Northwest winter wheat. *Crop Breeding, Genetics, and Genomics* 1: e190012.

Ly, D., Hamblin, M., Rabbi, I., Melaku, G., Bakare, M., Gauch, H. G., Okechukwu, R., Dixon, A. G. O., Kulakow, P. and Jannink, J.-L. 2013. Relatedness and genotype × environment interaction affect prediction accuracies in genomic selection: A study in Cassava. *Crop Sci.* 53: 1312–1325.

Meuwissen, T. H. E., Hayes, B. J. and Goddard, M. E. 2001. Prediction of total genetic value using genome-wide dense marker maps. *Genetics* 157: 1819–1829.

Michel, S., Ametz, C., Gungor, H., Akgöl, B. and Epure, D. 2017. Genomic assisted selection for enhancing line breeding: merging genomic and phenotypic selection in winter wheat breeding programs with preliminary yield trials. *Theor. Appl. Genet.* 130: 363–376.

Moser, G., Khatkar, M. S., Hayes, B. J. and Raadsma, H. W. 2010. Accuracy of direct genomic values in Holstein bulls and cows using subsets of SNPs. *Genet. Sel. Evol.* 42: 37.

Muleta, K. T., Bulli, P., Zhang, Z., Chen, X. and Pumphrey, M. 2017. Unlocking diversity in germplasm collections via genomic selection: a case study based on quantitative adult plant resistance to stripe rust in spring wheat. *The Plant Gen.* 10(3): 1–15.

Norman, A., Taylor, J., Edwards, J. and Kuchel, H. 2018. Optimising genomic selection in wheat: Effect of marker density, population size and population structure on prediction accuracy. *Genes Genom. Genet.* 8: 2889–2899.

Odilbekov, F., Armonienė, R., Koc, A., Svensson, J. and Chawade, A. 2019. GWAS assisted genomic prediction to predict resistance to Septoria Tritici Blotch in Nordic winter wheat at seedling stage. *Front Genet.* 10: 1224. doi:10.3389/fgene. 2019.01224.

Ornella, L., Singh, S., Pérez-Rodríguez, P., Burgueño, J., Singh, R., Tapia, E., Bhavani, S., Dreisigacker, S., Braun, H. J., Mathews, K. and Crossa, J. 2012. Genomic prediction of genetic values for resistance to wheat rusts. *Plant Gen.* 5: 136–148.

Pérez-Rodríguez, P., Crossa, J., Rutkoski, J., Poland, J., Singh, R., Legarra, A., Autrique, E., Campos, G., Burgueño, J. and Dreisigacker, S. 2017. Single-step genomic and pedigree genotype × environment interaction models for predicting wheat lines in international environments. *The Plant Gen.* 10.

Philipp, N., Weise, S., Oppermann, M., Borner, A., Andreas, G., Keilwagen, J., Kilian, B., Zhao, Y., Reif, J. C. and Schulthess, A. W. 2018. Lveraging the use of historical data gathered during seed regeneration of an *ex situ* genebank collection of wheat. *Front. Plant Sci.* 9: 609.

Poland, J. A., Brown, P. J., Sorrells, M. E. and Jannink, J. L. 2012. Development of high-density genetic maps for barley and wheat using a novel two-enzyme genotyping-by-sequencing approach. *PLoS One* 7: e32253.

Rutkoski, J., Poland, J., Singh, R., Huerta-Espino, J., Bhavani, S., Barbier, H., Rouse, M., Jannink, J. L. and Sorrells, M. 2014. Genomic selection for quantitative adult plant stem rust resistance in wheat. *The Plant Gen.* 7.

Rutkoski, J., Poland, J., Mondal, S., Autrique, E., Párez, L. G., Crossa, J., Reynolds, M. and Singh, R. 2016. Canopy temperature and vegetation indices from high-throughput phenotyping improve accuracy of pedigree and genomic selection for grain yield in wheat. *Genes Genet. Genom.* 6.

Saint Pierre, C., Burgueño, J., Crossa, J., Fuentes Davila, G., Figueroa López, P., Solis Moya, E., Ireta Moreno, J., Hernandez Muela, V. M., Zamora Villa, V. M., Vikram, P., Mathews, K., Sansaloni, C., Sehgal, D., Jarquin, D., Wenzl, P. and Singh, S. 2016. Genomic prediction models for grain yield of spring bread wheat in diverse agro-ecological zones. *Sci. Rep.* 6: 27312.

Sarinelli, J. M., Murphy, J. P., Tyagi, P., Holland, J. M., Johnson, J. W., Mergoum, M., Mason, R. E., Babar. A., Harrison, S., Sutton, R., Griffey, C. A. and Guedira, G. B. 2019. Training population selection and use of fixed effects to optimize genomic predictions in a historical USA winter wheat panel. *Theor. Appl. Genet.* 132: 1247–1261.

Sehgal, D., Autrique, E., Singh, R., Ellis, M., Singh, S. and S. Dreisigacker. 2017. Identification of genomic regions for grain yield and yield stability and their epistatic interactions. Sci. Rep. 7: 41578.

Sehgal, D., Rosyara, U., Mondal, S., Singh, R., Poland, J. and Dreisigacker, S. 2020. Incorporating genome-wide association mapping results into genomic prediction models for grain yield and yield stability in CIMMYT spring bread wheat. *Front. Plant Sci.* 11: 197.

Shanti, M. L., Shenoy, V. V., Devi, L. G., Kumar, M. V., Premalatha, P., Kumar, N. G., Shashidhar, H. E., Zehr, U. B. and Freeman, W. H. 2010. Marker-assisted breeding for resistance to bacterial leaf blight in popular cultivar and parental lines of hybrid rice. *J. Plant Pathol.* 92: 495–501.

Su, G., Christensen, O. F., Janss, L. and Lund, M. S. 2014. Comparison of genomic predictions using genomic relationship matrices built with different weighting factors to account for locus-specific variances. *J. Dairy Sci.* 97: 6547–6559.

Sun, J., Rutkoski, J. E., Poland, J. A., Crossa, J., Jannink, J. L. and Sorrells, M. E. 2017. Multitrait, random regression, or simple repeatability model in high-throughput phenotyping data improve genomic prediction for wheat grain yield. *Plant Gen.* 10(2).

Velu, G., Crossa, J., Singh, R. P., Hao, Y., Dreisigacker, S., Perez-Rodriguez, P. et al. 2016. Genomic prediction for grain zinc and iron concentrations in spring wheat. *Theor. Appl. Genet.* 129(8).

Walter, J., Edwards, J., Cai, J., McDonald, G., Miklavcic, S. J. and Kuchel, H. 2019. High-throughput field imaging and basic image analysis in a wheat breeding programme. *Front. Plant Sci.* 10: 449.

Wang, X., Li, L., Yang, Z., Zheng, X., Yu, S., Xu, C. and Hu, Z. 2016. Predicting rice hybrid performance using univariate and multivariate GBLUP models based on North Carolina mating design II. *Heredity* 118: 302.

Zhang, Z., Ober, U., Erbe, M., Zhang, H., Gao, N., He, J., Li, J. and Simianer, H. 2014. Improving the accuracy of whole genome prediction for complex traits using the results of genome wide association studies. *PLOS One* 9: e93017.

Zhang, Y. and Zhang, N. 2018. Imaging technologies for plant high-throughput phenotyping: A review. *Front. Agric. Sci. Eng.* 5(4): 406–419.

Zhao, X., Tan, G., Xing, Y., Wei, L., Chao, Q., Zuo, W., Lubberstedt, T. and Xu, M. 2012. Marker-assisted introgression of qHSR1 to improve maize resistance to head smut. *Mol. Breed.* 30: 1077–1088.

4

Genomic Selection in Rice: Current Status and Future Prospects

Anilkumar C,[1] *Rameswar Prasad Sah,*[1]
Muhammed Azharudheen TP,[1] *Sunitha NC,*[2] *Sasmita Behera,*[1]
BC Marndi,[1] *Tilak Raj Sharma*[3] and *Anil Kumar Singh*[3,4,*]

ABSTRACT

Genomic selection (GS) was first developed in animal systems as a modified version of marker-assisted selection. GS gained its popularity in plant breeding very quickly due to its effectiveness in selections. It makes use of genome-wide marker effects to assign genotypic values to individuals in the population from which they are selected for improving quantitative traits. Even though, the success of GS is evidenced by several reports in many crops, its implementation in rice is still limited. Understanding the basic principle and workflow of GS may encourage rice breeders to adapt it in the rice breeding programs. A sound knowledge on predictive accuracy and factors affecting it, application of different GS models under different breeding objectives is a prerequisite before implementing it in breeding programs. In this chapter, we discuss practical applications of GS in rice breeding, integrating components of GS with other breeding programs and accounting for genotypes by environment interaction (GEI) in GS assisted rice breeding programs.

[1] ICAR-National Rice Research Institute, Cuttack, Odisha, India.
[2] University of Agricultural Sciences, Bangalore, Karnataka, India.
[3] ICAR-Indian Institute of Agricultural Biotechnology, Ranchi, India.
[4] ICAR-National Institute for Plant Biotechnology, New Delhi, India.
* Corresponding author: anils13@gmail.com

Introduction

Over the years, the main objective of rice breeding programs has been to develop high yielding varieties to meet the global food demand and keep pace with an alarmingly increasing global population (Xu et al., 2014). Earlier methods of rice breeding were dependent on phenotype-based selection and evaluation of selected elite lines for the improvement of target traits. Advances in science and the invention of many cutting-edge technologies in the field of breeding shifted the focus towards using genomic resources for plant breeding. Evolution of DNA marker-based techniques and genomic approaches paved the way for accelerated advancements in rice improvement programs (Xu, 2010). However, the stagnated genetic gain per breeding cycle over the years in rice was not significantly altered. A breakthrough in improving the genetic gain has not been achieved with the available breeding methodologies (Muralidharan et al., 2019). Given the success of Genomic Selection (GS) to enhance the rate of genetic gain in livestock breeding, the method was introduced to plant breeding (Meuwissen et al., 2001; Schaeffer, 2006; Smith et al., 2008). GS was proposed as a promising genomic tool to address many challenges in crop breeding, including those in rice breeding.

GS has risen by virtue of new high-throughput DNA marker-based technologies and new statistical or data science advancements that analyze genetic frameworks of complex quantitative traits under well defined statistical models with accurate model effects. GS can be considered as a modified form of Marker Assisted Selection (MAS) with genome-wide dense DNA markers to predict genetic values with high accuracy to allow further selection (Hickey et al., 2017). However, GS is more suitable for quantitative traits under the control of a large number of small effect genes (He et al., 2016). The main goal of GS is to combine the phenotype of a training population with its genome-wide DNA marker effect and predict the genetic value of future individuals in breeding/test populations (Desta and Ortiz, 2014). The main components of GS are (i) utilizing genome-wide markers simultaneously to develop a genotype-phenotype relationship model in one population (called training population) accounting for genome-wide linkage disequilibrium (LD) among markers, and (ii) predicting the genomic estimated breeding values (GEBV) based on the model in future candidates of other related populations (called breeding population) (Meuwissen et al., 2001; Heffner et al., 2009). The success of GS depends on the accuracy of prediction and predictability of models adapted in different crops (Crossa et al., 2017). However, the predictability is affected by many factors including relatedness of the population, marker density, trait heritability and sample size (Heslot et al., 2012).

Rice being one of the most extensively researched crops among the cereals, its genomic resources have been developed to a greater extent. Most of the breeding methods developed till date have already been applied to this crop successfully. GS has also helped in achieving greater success in rice improvement, selecting parents based on breeding values for hybridization, GEBV based blast resistance source identification, hybrid breeding and varietal improvement, are some examples. Some of the efforts made to implement GS in rice breeding for improving yield and very few on disease resistance breeding are listed in Table 1. Most of the GS studies implemented in rice are focused to improve the prediction accuracy of different models

Table 1. Genomic selection studies conducted in rice.

Sl. No.	Target trait	Plant material	Statistical method	Reference
1	Yield, tillers and grain number	240 RILs and 360 derived crosses	LASSO, GBLUP and SSVS	Xu et al. (2014)
2	Plant height and protein content	Diversity panel of 413 accessions	GBLUP	Guo et al. (2014)
3	Flowering, plant height, panicle weight and yield	Synthetic population of size 400	GBLUP, rrBLUP and LASSO	Grenier et al. (2015)
4	8 agronomic traits including yield	110 Asian rice cultivars	GBLUP, RKHS and LASSO	Onogi et al. (2015)
5	Yield and related traits	369 elite breeding lines	GBLUP and RKHS	Spindel et al. (2015)
6	Yield and related traits	Structured diversity panel of 413 accessions	GBLUP	Isidro et al. (2015)
7	Heading dates	174 BILs along with parental lines	EBL and GBLUP	Onogi et al. (2016)
8	8 agronomic traits including grain yield	575 F_1 hybrids	GBLUP	Wang et al. (2017)
9	Grain yield and related traits	575 hybrids generated using NC II design	GBLUP, LASSO, SVM and PLS	Xu et al. (2018)
10	Grain filling ability	Diversity panel of 128 lines	GBLUP, PLS	Yabe et al. (2018)
11	Resistance to rice blast	161 African rice accessions	GBLUP	Huang et al. (2019)
12	Grain yield and plant height	Diversity panel of 280 accessions	GBLUP and RKHS	Bhandari et al. (2019)
13	Arsenic content in flag leaf and cargo grain	Diversity panel of 225 accessions and 95 elite lines	GBLUP and RKHS	Frouin et al. (2019)
14	Yield, grain weight and grain number	210 F_9 RILs	2D BLUP-HAT	Wang et al. (2020)
15	Yield and related traits	1495 hybrids	GBLUP	Yanru Cui et al. (2020)

RIL-Recombinant inbred line; BIL-Backcross in bred line; NC II-North Carolina mating design II; LASSO-Least Absolute Shrinkage and Selection Operator; GBLUP-genomic best linear unbiased prediction; SSVS-Stochastic search variable selection; rrBLUP-ridge regression best linear unbiased prediction; RKHS-reproducing kernel Hilbert space regression; PLS-partial least square; 2D BLUP-HAT-two dimensional bivariate BLUP-HAT; SVM-Vector support machine.

adapted for improvement of a few yield and related traits (Bernardo and Yu, 2007; Lorenz, 2013; Zhang et al., 2015).

Genetic and genomic resources for genomic selection in rice

Rice being a global crop, researched across the world, the highest number of genomic resources have been developed for it. All types of marker systems, starting from hybridization-based marker (RFLP and AFLP) to sequence based (SNP) markers have been employed and exploited to a greater extent. The availability of sequence information opened the avenue for functional genomic research in the last one decade. These NGS platforms along with bioinformatic pipelines offered practical toolkits for helping high through-put identification of genes and their effects in the expression of phenotypes. These advances in rice molecular genetics, phenomics and genomics offered faster development of potential breeding tools. Among these, MAS has offered greater advantages with a small number of markers, in tracking and tagging markers to genes of interest, but its efficiency is limited in precision breeding (Jena and Mackill, 2008). MAS has been successfully employed in introgression of several major genes or QTL regions controlling important phenotypes in rice. However, it is difficult to elucidate the combinations of genes with minor effects in generating novel phenotypes in rice, which is a limitation of MAS (Bernardo, 2010; Fujino et al., 2019). Utilizing sequence-based markers, specifically SNPs in identifying the high performing progeny based on breeding values replaces the progeny testing and saves generation time. Marker assisted and genomic selection work on same basic principle of marker-based selection, but both the tools differ in many aspects (Table 2).

The success of GS mainly depends on the prediction accuracy which is also referred to as the correlation between GEBV and true breeding value. In practice, the prediction accuracy is examined by correlating GEBV with the realized phenotype. To improve the efficiency of GS model training, rice breeding requires models that account for a variety of situations, primarily to account for genotype-environment interactions (GEI) and to manage genotype screening trials data over several environments and years in assessing tolerance to biotic and abiotic stresses.

Factors affecting predictive accuracy of genomic selection in rice

Selection response is the most important factor in the breeder's equation, along with trait heritability and time required to complete the selection cycle; increasing the selection response can increase the genetic gain of the target trait. Many simulation and empirical experiments have shown that the GS can increase genetic gain by reducing the breeding cycle and increasing selection intensity (Crossa et al., 2014). However, the accuracy of prediction is affected

Table 2. Differences between marker assisted selection and genomic selection.

Marker assisted selection	**Genomic selection**
• It is an indirect selection of a difficult to phenotype trait using genetic markers linked to the genomic region controlling the trait of interest.	• It is a variant of MAS based on genomic estimated breeding values (GEBVs) estimated using all the markers' effects using a trained GS model.
• It is effective for traits with major effects, i.e., oligogenic traits/major QTLs.	• It is effective for traits with small effects along with major effects, i.e., polygenic traits/major and minor QTLs.
• Detection/mapping and validation of markers linked to QTL associated with the trait of interest is prerequisite to implement MAS.	• Training a suitable GS model using phenotypic and genotypic data in a training population is a pre-requisite to implement GS.
• It is a targeted approach where only markers linked to few validated major QTL are used to implement MAS.	• It is a wholistic approach where all the markers used in training a GS model are used to implement GS in the breeding population.
• It is applied on any population of a given crop if the QTL is validated, which is very rare.	• It is workable on a breeding population which is related to/derivative of the training population.
• Accurate implementation of MAS is achieved if major QTL with tightly linked markers validated across genetic backgrounds are employed. Hence, fine mapping and validation of QTL influence accuracy of MAS.	• Prediction accuracy can be improved by optimizing factors like number of markers, prediction model, population size & structure of training population, relationship between TP and BP, size ratio of number of TP to BP, LD status of population and mor.
• Relatively less effective for improving quantitative traits as only a few genomic regions associated with the trait are considered.	• Highly effective in improving quantitative traits as this approach considers effects of all the markers across the genome and the GS model can be further improvised considering environmental trials.
• MAS could be implemented to complement any of the conventional breeding strategies like MA-Backcross, MA-Pedigree, MA-Recurrent selection.	• Genomic selection is more appropriately implemented in line development breeding. Of late, it is also being used extensively in hybrid breeding.
• Genetic gain per unit time is less, i.e., much time is spent on QTL detection and validation.	• Genetic gain per unit time is relatively high as all the QTLs with major and minor effects are considered.
• It has limitations of linkage drag, background noise and environmental instability especially for a quantitative trait.	• It is limited by factors influencing prediction accuracy.

by many factors (Fig. 1), such as population type, population size, genotyping method and genome coverage, marker type, target trait, heritability, statistical model used to estimate GEBVs, type of breeding scheme and more (Table 3). To achieve maximum benefits in a GS assisted breeding program, all factors affecting the success of GS should be considered (Xu et al., 2020).

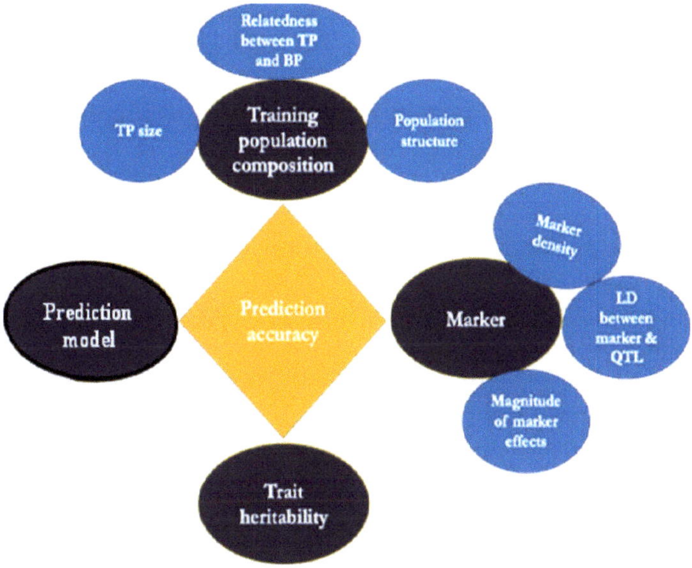

Fig. 1. Pictorial representation of factors affecting predictive accuracy in genomic selection.

Table 3. Factors affecting prediction accuracy of GS in rice breeding.

Component	Contributors
Population	Population type, population size, selection proportion, selection method, novel germplasm introduction, population structure
Marker	Molecular markers used, marker types, marker density, marker distribution, target genes, LD between marker and QTL
Trait heritability	Trait heritability Estimation method, number of genes controlling the trait, management at field level, environmental interactions, mating design used, precision of phenotyping
GS model	Model used, genetic effects estimated, single model or multi-model approach, accuracy of model estimates
Breeding schemes	Breeding program design, selection scheme, cost of breeding program, integrated breeding platforms adapted

In essence, these factors solely may not affect the predictive accuracy of GS to a greater extent, when they are congruent, the results indicate practical deviations in the output of GS. Hence, deep understanding of each factor affecting success of GS is the need of the hour for better planning of GS assisted breeding in rice.

Effect of training and breeding populations on prediction accuracy

Phenotypic and genotypic data of the training population is used to build a genomic prediction model, which is used to calculate GEBV of the genotyped breeding population. Hence, accuracy of prediction depends on the composition of the training population (size and genetic structure), markers (number), relationship between the training population (TP) and the breeding population (BP), genetic architecture of the trait and, the model used. These factors are interconnected and are relative (Grenier et al., 2015). There are several reports on the effect of each of the factors affecting prediction accuracy using cross validation tests. However, conclusions of these findings collectively help optimization of these factors to achieve higher prediction accuracy.

Phenotypic and genotypic data of the training population are considered to build the GS model. Accuracy of prediction is reported to be affected by (a) size of the TP, (b) complexity of genetic structure and/or diversity (between sub-populations) in the TP and (c) relationship between the training and breeding population. Several early reports opined an increase in prediction accuracy with an increase in TP size (Lorenzana and Bernardo, 2009; Asoro et al., 2011). However, a TP size of around 100–500 genotypes has been observed to be optimum in rice (Bhandari et al., 2019; Hassen et al., 2017; 2018). The optimum size of the training population depends on genetic structure and trait heritability. High prediction accuracy can be achieved for high heritable traits even with a moderate population size (Spindel et al., 2015). On the contrary, highly variable TP of ~ 50 individuals yielded low prediction accuracy even for a trait with high heritability (Onogi et al., 2015). Designing of training population through suitable sampling of the genetic variation/diversity in it is necessary to correlate its genotype and phenotype for precise genomic prediction modelling (Spindel and Iwata, 2018; Akdemir et al., 2019). The relationship between training and validation/breeding population is another factor influencing prediction accuracy. Higher the relatedness between TP and BP, higher is the prediction accuracy (Desta and Ortiz, 2014). Composing TP and BP using full-sibs showed higher accuracy than using half-sib families owing to higher relatedness in full-sibs (Guo et al., 2012; Riedelsheimer et al., 2013). For a highly heritable trait, Lehermeier et al. (2014) showed 50 full sibs were sufficient to achieve a predictive power of 375–675 half-sibs. Population structure, relatedness and/or diversity in the training population affect the composition of the training and breeding populations, thereby affecting prediction accuracy (Guo et al., 2014; Grenier et al., 2015; Hassen et al., 2017). In cross-validation sets, stratified sampling ensuring representatives from each of the sub-populations in each cross-validation fold outperformed random sampling (Guo et al., 2014; Isidro et al., 2015). Similar findings were reported by Grenier et al. (2015) and Hassen et al. (2017) where high accuracy was achieved when progeny prediction was based on a model trained on the TP with related individuals. Optimization of size considering genetic structure and relationship between TP and BP is more successful. An increase in the prediction accuracy was reported with an increase in the ratio of population sizes of the training and breeding populations (Hassen et al., 2017). Liu et al. (2018) reported a ratio of three as optimum beyond which prediction accuracy was negligible. With the inclusion of the environment component for prediction, sampling or choice of TP individuals stable across environments improved prediction accuracy (Hoffstetter et al., 2016; Hassen et al., 2018). Similar improvement in prediction accuracy by composing TP with lines specific to both wet and dry seasons were reported in rice breeding lines from IRRI (Spindel et al., 2015).

Effect of genotypic data on prediction accuracy

Marker density used for genomic prediction depends on the genetic structure of TP, LD status and trait heritability. Extensive genomic resources (markers) have been developed post-sequencing of a rice genome, which was further refined following the advent of next generation sequencing (NGS). A large number of markers irrespective of whether SSR or SNP (majority of GS studies used SNP), covering the whole genome are preferred. This is based on assumption that always some markers would be in LD with any QTL. Prediction accuracy has been reported to increase with increase in marker density only when trait heritability is high (Bernardo and Yu, 2007; Heffner et al., 2011; Wang et al., 2014; Arruda et al., 2016; Zhang et al., 2016). However, the prediction accuracy was found to plateau at certain densities of markers beyond which there was no increase in it (Xu et al., 2020). Bi-parental populations like RIL or DH with a narrow genetic base always have high LD between markers and QTL (Wang et al., 2018). Hence, a larger number of markers have been found ideal for a natural population/diverse germplasm collection with a complex genetic structure than for bi-parental populations (Liu et al., 2018). Spindel et al. (2015) reported that the number of markers > 7500 (~ 19 SNP per Mb) did not improve prediction accuracy among subsets of 73,147 SNPs on the population of 363 Indica lines. Similarly, Grenier et al. (2015) and Bhandari et al. (2019) concluded 13 SNPs and 27 SNPs per Mb were sufficient to improve prediction accuracies, respectively. These differential reports on optimum marker density could be attributed to differences in LD reinforcing lower marker densities attributable to large LD ($r^2 \geq 0.50$). Few studies reported that prediction accuracy is affected by systematic selection or elimination of markers based on their effects (Hoffstetter et al., 2016; Schulz-Streeck et al., 2013), but no such studies on rice have been conducted so far.

Effect of trait characteristics on prediction accuracy

The genomic prediction accuracy in rice breeding programs depends highly on trait heritability along with marker density and population size. Since, rice is a self-pollinated crop with huge LD blocks that are passed down through generations, many agronomic traits maintain high heritability. Many empirical studies revealed higher heritability values for days to flowering and plant height characters than grain yield (Grenier et al., 2015). Poor predictive accuracy was recorded for traits having lower heritability in a panel of 413 diverse accessions (Guo et al., 2015). Spindel et al. (2015) reported lower heritability for grain yield compared to days to flowering and respective poor predictive accuracy for grain yield than days to flowering. Utilizing narrow sense heritability for estimating breeding values in rice is more appropriate than considering broad sense heritability for achieving rapid genetic gains (Nakaya and Isobe, 2012). For breeding programs dealing with traits with low heritability, the breeder should be more specific towards using high-density genotyping platforms or opt to pool multiple populations that increase allele diversity while designing training populations to increase the prediction accuracy of GEBVs.

Effect of prediction methods on predictive accuracy

The genomic selection methods rely on the selection methods used to find genetic factors underlying the expression and distribution of phenotypes. Before adopting a particular GS model in a breeding program, it is important to test the effectiveness of the model for traits under study in order to ensure a better fit of the model to achieve the maximum benefit of genetic gain. The predictive models can be upgraded and replaced depending on the requirement. Several GS models are developed and used with different software packages (Table 4); most of them are developed under R software background and only few with a python background. Atleast more than ten models have been employed in rice across different empirical GS studies.

Few researchers have compared the effectiveness of models, for, e.g., Onogi et al. (2015) compared nine GS models using a panel of 110 rice genotypes with eight agronomic characters and inferred that GBLUP and rrBLUP models are most accurate for a single trait, and RKHS model for multiple traits. A similar kind of interpretation was also made by Greneir et al. (2015) with additional information on using a random forest model for multiple traits to increase predictive ability. The Bayesian lasso method and its variants showed a stable performance compared to the other methods in the study conducted by Spindel et al. (2015). The basic GS models working on the principle of regression assume a linear relationship between phenotypic trait expressions and genome-wide markers. Therefore, these models capture additive effects of markers and are much more suitable to predict phenotypic traits in self-pollinated species like rice, however, for GS assisted hybrid rice breeding programs, utilizing the models capturing the dominance and epistasis effects are recommended (Martini et al., 2017).

Table 4. List of software packages used for genomic selection in breeding programmes (Tong and Nikoloski, 2021).

Software/package	Description	Reference
rrBLUP	ridge regression Best Linear Unbiased Prediction	Endelman (2011)
BLR	Bayesian linear regression	P′erez et al. (2010)
BGLR	Bayesian generalized linear regression	P′erez and de los Campos (2014)
GS3	GS, Gibbs sampling, Gauss Seidel	Legarra et al. (2016)
solGS	web-based tool for GS	Tecle et al. (2014)
GVCBLUP	genomic prediction of additive and dominance effects	Wang et al. (2014)
GenoMatrix	pedigree-based genomic prediction	Nazarian and Gezan (2016)
ShinyGPAS	interactive genomic prediction	Morota (2017)
Gselection	feature select and genomic prediction	Majumdar et al. (2019)
GenomicLand	genome-wide association study and genomic prediction	Azevedo et al. (2019)
SeqBreed	Python tool for genomic prediction	P′erez-Enciso et al. (2020)
GVCHAP	genomic prediction using haplotypes and SNPs	Prakapenka et al. (2020)
BWGS	GS in wheat breeding program	Charmet et al. (2020)
MTGS	GS using multiple traits	Budhlakoti et al. (2019)
BMTME	Bayesian multi-trait and multi-environment model	Montesinos-L′opez et al. (2019)
BGGE	Bayesian genomic genotype × environment interaction	Granato et al. (2018)

Accounting GEI in GS assisted rice breeding

Genotype × environment interactions (GEI) play a major role in the expression of complex traits. Change in phenotypic expression of genotypes evaluated under different environmental conditions is considered to account for GEI (Cooper and Hammer, 1996). Several models to account for GEI have appeared in the recent era as a result of advances in data science; Burgueo et al. (2012) were the first to report the expansion of GBLUP to multi-environment trials to account for GEI. In rice, Hassen et al. (2018) implemented multi-environment genomic predictions to manage abiotic stress experiments. They also compared prediction accuracy under a single environment and multiple environments with two cross validation strategies. The extended GBLUP model (Lopez-Cruz et al., 2015) and the extended RKHS model (Cuevas et al., 2017), which integrate environmental effects inferred that models with multi-environment data deliver higher predictive accuracy than a singleenvironment. Availability of literature in the field of rice research for application of these tools is limited; however, deep learning and machine learning tools were employed in predicting yield levels in barley in combination with historical weather data that can be extended to rice GS breeding programs (Gillberg et al., 2019). Considering G × E, developing GS models is still a challenge, few reports are available till date and more research is required to untangle the effects of the environment on different genotypes and reliably predict their plasticity.

Conclusion

Under the changing climatic conditions, rice breeders need to develop promising varieties at an accelerated rate. To increase the genetic gain per breeding cycle and per unit cost, adapting new precision breeding approaches is more relevant. Advances in genomics technologies and data science offered promising tools to speed up the breeding activities with effective genetic gains per generation. One such tool is genomic selection, which made significant transformations in maize and wheat breeding across the globe. GS may be considered as a potent and valuable approach for rice breeding in the future. GS helps with more precise conversion of genotype value into phenotype compared to MAS. This idea may help to design ideal genotypes without any QTL mapping and pyramiding of QTLs. Integrating environmental variables to GS models will also account for higher levels of precision in genotype prediction. However, GS is not a solution to the drawbacks of MAS but helps to upgrade MAS to the next phase through GEBV based selection of complex traits in rice breeding programs. It will be more fruitful if GS is integrated into rice breeding programs with appropriate standard operating procedures (SOP) to improve complex traits in the near future.

References

Akdemir, D., Beavis, W., Fritsche-Neto, R., Singh, A. K., Isidro-Sánchez, J. et al. 2019. Multi-objective optimized genomic breeding strategies for sustainable food improvement. *Heredity* 122(5): 672–683.

Arruda, M. P., Lipka, A. E., Brown, P. J., Krill, A. M., Thurber, C. et al. 2016. Comparing genomic selection and marker-assisted selection for Fusarium head blight resistance in wheat (*Triticum aestivum* L.). *Molecular Breeding* 36(7): 1–11.

Asoro, F. G., Newell, M. A., Beavis, W. D., Scott, M. P., Jannink, J. L. et al. 2011. Accuracy and training population design for genomic selection on quantitative traits in elite North American oats. *The Plant Genome* 4(2).

Azevedo, C. F., Nascimento, M., Fontes, V. C., de Silva, F. F., de Resende, M. D. V. et al. 2019. Genomic Land: software for genome-wide association studies and genomic prediction. *Acta Sci. Agron.* 41: e45361. https://doi.org/10.4025/actasciagron. v41i1.45361.

Bernardo, R. 2010. Genomewide selection with minimal crossing in self-pollinated crops. *Crop Science* 50: 624–627.

Bernardo, R. and Yu, J. 2007. Prospects for genome-wide selection for quantitative traits in maize. *Crop Science* 47(3): 1082–1090.

Bhandari, A., Bartholomé, J., Cao-Hamadoun, T. V., Kumari, N., Frouin, J. et al. 2019. Selection of trait-specific markers and multi-environment models improve genomic predictive ability in rice. *Plos One* 14(5): p.e0208871.

Budhlakoti, M., Mishra, D. C. M. and Rai, A. 2019. R Package MTGS: Genomic Selection Using Multiple Traits. https://CRAN.R-project.org/package=MTGS.

Burgueño, J., de los Campos, G., Weigel, K. and Crossa, J. 2012. Genomic prediction of breeding values when modelling genotype × environment interaction using pedigree and dense molecular markers. *Crop Science* 52(2): 707–719.

Charmet, G., Tran, L. G., Auzanneau, J., Rincent, R. and Bouchet, S. 2020. BWGS: a R package for genomic selection and its application to a wheat breeding programme. *PLoS One* 15: e0222733. https://doi.org/10.1371/journal.pone.0222733.

Cooper, M. and Hammer, G. L. 1996. *Plant Adaptation and Crop Improvement.* CAB International, Wallingford, UK.

Crossa, J. et al. 2014. Genomic prediction in CIMMYT maize and wheat breeding programs. *Heredity* 112: 48–60.

Crossa, J. et al. 2017. Genomic selection in plant breeding: methods, models, and perspectives. *Trends Plant Sci.* 22: 961–975.

Cuevas, J., Crossa, J., Montesinos-López, O. A., Burgueño, J., Pérez-Rodríguez, P. et al. 2017. Bayesian genomic prediction with genotype × environment interaction kernel models. *Genes, Genomes, Genetics* 7(1): 41–53.

Desta, Z. A. and Ortiz, R. 2014. Genomic selection: genome-wide prediction in plant improvement. *Trends in Plant Science* 19(9): 592–601.

Endelman, J. B. 2011. Ridge regression and other kernels for genomic selection with r package rrBLUP. *Plant Genome* 4: 250–255.

Frouin, J., Labeyrie, A., Boisnard, A., Sacchi, G. A., Ahmadi, N. et al. 2019. Genomic prediction offers the most effective marker assisted breeding approach for ability to prevent arsenic accumulation in rice grains. *PLoS ONE* 14(6): e0217516. DOI: 10.1371/journal.pone.0217516.

Fujino, K., Hirayama, Y. and Kaji, R. 2019. Marker-assisted selection in rice breeding programs in Hokkaido. *Breeding Science* 69(3): 383–392. https://doi.org/10.1270/jsbbs.19062.

Gillberg, J., Marttinen, P., Mamitsuka, H., Kaski, S., Stegle, O. et al. 2019. Modelling G3E with historical weather information improves genomic prediction in new environments. *Bioinformatics* 35: 4045–4052. https://doi.org/10.1093/bioinformatics/btz197.

Granato, I., Cuevas, J., Luna-V´azquez, F., Crossa, J., Montesinos-López, O. et al. 2018. BGGE: a new package for genomic-enabled prediction incorporating genotype × environment interaction models. *G3 Genes Genomes Genet.* 8: 3039–3047. https://doi.org/10.1534/g3.118.200435.

Grenier, C., Cao, T. V., Ospina, Y., Quintero, C., Châtel, M. H. et al. 2015. Accuracy of genomic selection in a rice synthetic population developed for recurrent selection breeding. *PloS One* 10(8): e0136594.

Guo, Z., Tucker, D. M., Lu, J., Kishore, V. and Gay, G. 2012. Evaluation of genome-wide selection efficiency in maize nested association mapping populations. *Theoretical and Applied Genetics* 124(2): 261–275.

Guo, Z., Tucker, D. M., Basten, C. J., Gandhi, H., Ersoz, E. et al. 2014. The impact of population structure on genomic prediction in stratified populations. *Theoretical and Applied Genetics* 127: 749–762. DOI:10.1007/s00122-013-2255-x.

Hassen, M., Bartholomé, J., Valè, G., Cao, T. V. and Ahmadi, N. 2018. Genomic prediction accounting for genotype by environment interaction offers an effective framework for breeding simultaneously for adaptation to an abiotic stress and performance under normal cropping conditions in rice. *Genes, Genomes, Genomics* 8(9): 2319–2332. DOI: 10.1534/g3.118.200098.

Hassen, M., Cao, T. V., Bartholomé, J., Orasen, G., Colombi, C. et al. 2017. Rice diversity panel provides accurate genomic predictions for complex traits in the progenies of biparental crosses involving members of the panel. *Theoretical and Applied Genetics* 131(2): 417–435. DOI: 10.1007/s00122-017- 3011-4.

He, S., Schulthess, A. W., Mirdita, V. et al. 2016. Genomic selection in a commercial winter wheat population. *Theor. Appl. Genet.* 129: 641–651. doi:10.1007/s00122-015-2655-1.

Heffner, E. L., Sorrells, M. E. and Jannink, J. L. 2009. Genomic selection for crop improvement. *Crop Science* 49(1): 1–12.

Heffner, E. L., Jannink, J. L. and Sorrells, M. E. 2011. Genomic selection accuracy using multifamily prediction models in a wheat breeding program. *The Plant Genome* 4(1).

Heslot, N., Sorrells, M. E., Jannink, J. L. and Yang, H. P. 2012. Genomic selection in plant breeding: A comparison of models. *Crop Science* 52: 146–160.

Hickey, J. M., Chiurugwi, T., Mackay, I., Powell, W., Implementing Genomic Selection in CBPWP. 2017. Genomic prediction unifies animal and plant breeding programs to form platforms for biological discovery. *Nat. Genet.* 49: 1297–1303.

Hoffstetter, A., Cabrera, A., Huang, M. and Sneller, C. 2016. Optimizing training population data and validation of genomic selection for economic traits in soft winter wheat. *G3: Genes, Genomes, Genetics* 6(9): 2919–2928.

Huang, M., Balimponya, E. G., Mgonja, E. M., McHale, L. K., Luzi-Kihupi, A. et al. 2019. Use of genomic selection in breeding rice (*Oryza sativa* L.) for resistance to rice blast (*Magnaporthe oryzae*). *Molecular Breeding* 39(8): 114.

Isidro, J., Jannink, J. L., Akdemir, D., Poland, J., Heslot, N. et al. 2015. Training set optimization underpopulation structure in genomic selection. *Theoretical and Applied Genetics* 128: 145–158. DOI: 10.1007/s00122-014-2411-y.

Jena, K. K. and Mackill, D. J. 2008. Molecular markers and their use in marker-assisted selection in rice. *Crop Science* 48: 1266–1276.

Legarra, A., Ricard, A. and Filangi, O. 2016. GS3: Genomic Selection-Gibbs Sampling-Gauss Seidel. https://github.com/alegarra/gs3.

Lehermeier, C., Krämer, N., Bauer, E., Bauland, C., Camisan, C. et al. 2014. Usefulness of multiparental populations of maize (*Zea mays* L.) for genome-based prediction. *Genetics* 198(1): 3–16.

Liu, Y., Lu, S., Liu, F., Shao, C., Zhou, Q. et al. 2018. Genomic selection using BayesCπ and GBLUP for resistance against *Edwardsiella tarda* in Japanese flounder (*Paralichthys olivaceus*). *Marine Biotechnology* 20(5): 559–565.

Lopez-Cruz, M., Crossa, J., Bonnett, D., Dreisigacker, S., Poland, J. et al. 2015. Increased prediction accuracy in wheat breeding trials using a marker × environment interaction genomic selection model. *Genes, Genomes, Genetics* 5(4): 569–582.

Lorenz, A. J. 2013. Resource allocation for maximizing prediction accuracy and genetic gain of genomic selection in plant breeding: a simulation experiment. G3(3): 481–491. doi:10.1534/ g3.112.004911.

Lorenzana, R. E. and Bernardo, R. 2009. Accuracy of genotypic value predictions for marker-based selection in biparental plant populations. *Theoretical and Applied Genetics* 120(1): 151–161.

Majumdar, S. G., Rai, A. and Mishra, D. C. 2019. R Package GSelection: Genomic Selection. https://cran.r-project.org/web/packages/GSelection/index.html.

Martini, J. W. R., Gao, N., Cardoso, D. F., Wimmer, V., Erbe, M. et al. 2017. Genomic prediction with epistasis models: on the marker-coding-dependent performance of the extended

GBLUP and properties of the categorical epistasis model (CE). *BMC Bioinformatics* 18: 3. https://doi.org/ 10.1186/s12859-016-1439-1.

Meuwissen, T. H. E., Hayes, B. J. and Goddard, M. E. 2001. Prediction of total genetic value using genome-wide dense marker maps. *Genetics* 157: 1819–1829.

Montesinos-López, O. A., Montesinos-López, A., Luna-V´azquez, F. J., Toledo, F. H., P´erez-Rodríguez, P. et al. 2019. An R package for Bayesian analysis of multi-environment and multi-trait multi-environment data for genome-based prediction. *G3 Genes Genomes Genet.* 9: 1355–1369. https://doi.org/10.1534/ g3.119.400126.

Morota, G. 2017. ShinyGPAS: interactive genomic prediction accuracy simulator based on deterministic formulas. *Genet. Sel. Evol.* 49: 91. https://doi.org/10.1186/s12711-017-0368-4.

Muralidharan, K., Prasad, G. S., Rao, C. S. and Siddiq, E. A. 2019. Genetic gain for yield in rice breeding and rice production in India to meet with the demand from increased human population. *Curr. Sci.* 116(4): 544.

Nakaya, A. and Isobe, S. N. 2012. Will genomic selection be a practical method for plant breeding? *Annals of Botany* 110(6): 1303–1316.

Nazarian, A. and Gezan, S. A. 2016. GenoMatrix: a software package for pedigree-based and genomic prediction analyses on complex traits. *J. Hered.* 107: 372–379. https://doi.org/10.1093/jhered/esw020.

Onogi, A., Ideta, O., Inoshita, Y., Ebana, K., Yoshioka, T., Yamasaki, M. et al. 2015. Exploring the areas of applicability of whole-genome prediction methods for Asian rice (*Oryza sativa* L.). *Theoretical and Applied Genetics* 128(1): 41–53.

Onogi, A., Watanabe, M., Mochizuki, T., Hayashi, T., Nakagawa, H. et al. 2016. Toward integration of genomic selection with crop modelling: the development of an integrated approach to predicting rice heading dates. *Theoretical and Applied Genetics* 129(4): 805–817.

Pérez, P., de los Campos, G., Crossa, J. and Gianola, D. 2010. Genomic-enabled prediction based on molecular markers and pedigree using the bayesian linear regression package in r. *Plant Genome* 3: 106–116. https://doi.org/10.3835/ plantgenome2010.04.0005.

Pérez, P. and de los Campos, G. 2014. Genome-wide regression and prediction with the BGLR statistical package. *Genetics* 198: 483–495. https://doi.org/10.1534/ genetics.114.164442.

Pérez-Enciso, M., Ramírez-Ayala, L. C. and Zingaretti, L. M. 2020. SeqBreed: a python tool to evaluate genomic prediction in complex scenarios. *Genet. Sel. Evol.* 52: 7. https:// doi.org/10.1186/s12711-020-0530-2.

Prakapenka, D., Wang, C., Liang, Z., Bian, C., Tan, C. et al. 2020. GVCHAP: a computing pipeline for genomic prediction and variance component estimation using haplotypes and SNP markers. *Front. Genet.* 11: 282. https://doi.org/10.3389/fgene.2020.00282.

Riedelsheimer, C., Endelman, J. B., Stange, M., Sorrells, M. E., Jannink, J. L. et al. 2013. Genomic predictability of interconnected biparental maize populations. *Genetics* 194(2): 493–503.

Schaeffer, L. R. 2006. Strategy for applying genome-wide selection in dairy cattle. *Journal of Animal Breeding and Genetics* 123: 218–223.

Schulz-Streeck, T., Ogutu, J. O. and Piepho, H. P. 2013. Comparisons of single-stage and two-stage approaches to genomic selection. *Theoretical and Applied Genetics* 126(1): 69–82.

Smith, J. S., Hussain, T., Jones, E. S., Graham, G., Podlich, D. et al. 2008. Use of doubled haploids in maize breeding: implications for intellectual property protection and genetic diversity in hybrid crops. *Molecular Breeding* 22(1): 51–9.

Spindel, J., Begum, H., Virk, P., Collard, B., Redoña, E. et al. 2015. Genomic selection and association mapping in rice (*Oryza sativa*): Effect of trait genetic architecture, training population composition, marker number and statistical model on accuracy of rice genomic selection in elite, tropical rice breeding lines. *PLoS Genetics* 11: e1004982. DOI: 10.1371/ journal.pgen.1004982.

Spindel, J. and Iwata, H. 2018. Genomic selection in rice breeding. pp. 473–496. *In*: *Rice Genomics, Genetics and Breeding*. Springer, Singapore.

Tecle, I. Y., Edwards, J. D., Menda, N., Egesi, C., Rabbi, I. Y. et al. 2014. solGS: a web-based tool for genomic selection. *BMC Bioinformatics* 15: 398. https://doi.org/10.1186/s12859-014-0398-7.

Tong, H. and Nikoloski, Z. 2021. Machine learning approaches for crop improvement: Leveraging phenotypic and genotypic big data. *Journal of Plant Physiology* 257: p.153354.

Wang, C., Prakapenka, D., Wang, S., Pulugurta, S., Runesha, H. B. et al. 2014. GVCBLUP: a computer package for genomic prediction and variance component estimation of additive and dominance effects. *BMC Bioinformatics* 15: 270. https:// doi.org/10.1186/1471-2105-15-270.

Wang, S., Xu, Y., Qu, H., Cui, Y., Li, R. et al. 2020. Boosting predictabilities of agronomic traits in rice using bivariate genomic selection. *Briefings in Bioinformatics*.

Wang, X., Li, L., Yang, Z., Zheng, X., Yu, S. et al. 2017. Predicting rice hybrid performance using univariate and multivariate GBLUP models based on North Carolina mating design II. *Heredity* 118: 302–310.

Wang, X., Xu, Y., Hu, Z. and Xu, C. 2018. Genomic selection methods for crop improvement: Current status and prospects. *The Crop Journal* 6(4): 330–340.

Xu, S., Zhu, D. and Zhang, Q. 2014. Predicting hybrid performance in rice using genomic best linear unbiased prediction. *Proceedings of the National Academy of Sciences* 111(34): 12456–12461.

Xu, Y. 2010. *Molecular Plant Breeding* (Wallingford, UK: CABI Publishing).

Xu, Y., Wang, X., Ding, X. et al. 2018. Genomic selection of agronomic traits in hybrid rice using an NCII population. *Rice* 11: 32. https://doi.org/10.1186/s12284-018-0223-4.

Xu, Y., Liu, X., Fu, J., Wang, H., Wang, J. et al. 2020. Enhancing genetic gain through genomic selection: from livestock to plants. *Plant Communications* 1(1): 100005.

Yabe, S., Yoshida, H., Kajiya-Kanegae, H., Yamasaki, M., Iwata, H. et al. 2018. Description of grain weight distribution leading to genomic selection for grain-filling characteristics in rice. *PLoS ONE* 13(11): e0207627. https://doi.org/10.1371/journal.pone.0207627.

Yanru Cui, Li, R., Li, G., Zhang, F., Zhu, T. et al. 2020. Hybrid breeding of rice via genomic selection. *Plant Biotechnology Journal* 18(1): 57–67.

Zhang, J., Song, Q., Cregan, P. B. and Jiang, G. L. 2016. Genome-wide association study, genomic prediction and marker-assisted selection for seed weight in soybean (*Glycine max*). *Theoretical and Applied Genetics* 129(1): 117–130.

Zhang, X., Pérez-Rodríguez, P., Semagn, K. et al. 2015. Genomic prediction in biparental tropical maize populations in water-stressed and well-watered environments using low-density and GBS SNPs. *Heredity* 114: 291–299. doi:10.1038/hdy.2014.99.

5

Genomic Selection in Groundnut, Chickpea, and Pearl Millet: Applications and Prospects

Murali T Variath,[*,1] *Sunil Chaudhari,*[1] *Srinivas Samineni,*[2] *Dnyaneshwar B Deshmukh,*[1] *Anand Kannati*[3] and *Sudarshan Patil*[3]

ABSTRACT

Chickpea, groundnut, and pearl millet are three important climate smart crops of the semi-arid tropics with their ability to cope with slightly harsh temperature and soil moisture-deficit stress conditions to provide diversified nutritious diets to consumers. The last decade has witnessed the development of different molecular marker systems leading to the construction of genetic maps, identification of molecular markers linked to genes/quantitative trait loci for desired traits and their use in breeding programs. In recent years the progress in whole genome sequencing techniques has facilitated the use of new genomics assisted breeding tools such as genomic selection (GS) for improving complex traits in crops. The advantage in using genomic selection over other genomics assisted breeding methods, is that it considers all the major and minor alleles effects to predict genomic estimated breeding values that enhance selection intensity as well as accuracy and have

[1] Groundnut Breeding – Asia Program, International Crops Research Institute for the Semi-Arid Tropics (ICRISAT), Patancheru, Hyderabad, India 502324.
[2] Chickpea Breeding – Asia Program, ICRISAT, Patancheru, Hyderabad, India 502324.
[3] Pearl Millet Breeding – Asia Program, ICRISAT, Patancheru, Hyderabad, India 502324.
* Corresponding author: murali.tv2006@gmail.com

tremendous potential to accelerate the rate of genetic gains especially for complex quantitative traits. In all three crops, development and validation of GS models are in progress. The current chapter provides insights into the development of GS training populations, case studies of GS for target traits and GS prospects in all three crops.

Introduction

The production of adequate food to meet the demands of an enormously growing human population is a major challenge globally. Prediction statistics indicate 50% rise in the world population by 2050 (Tester and Langridge, 2010) which necessitates a further 70% increase in crop production (Furbank and Tester, 2011). These future demands and production predictions are further accentuated by several factors such as climate change, diminution of water and land resources, and poor soil/air/water health. The development of new crop varieties at an accelerated rate is needed to increase production as well as to meet the emerging challenges of biotic and abiotic stresses in a sustainable way.

Traditional breeding methods have been successful in transferring important genes from locally adapted germplasms and/or wild species into cultivated varieties to improve one or few closely related traits to improve yield potential. They rely mainly on crossing the plants and selecting individuals superior for morphological, physiological, agronomical and nutritional parameters than an existing variety or on introgression of specific traits that are lacking in the elite varieties. Several breeding approaches and selection methods such as selection index (SI) (Hazel and Lush, 1942) and best linear unbiased predictions (BLUP) (Henderson, 1975) have been used by breeders in the past to enhance accuracies of phenotypic selection to some extent while dealing with multiple traits together. Since the beginning of the 1990s, the development of molecular markers and advanced genomic tools has revealed the presence of widespread genetic variation in the genomes. This technology has enabled breeders to capture the genetic variation and selected target traits in breeding programs through genomics assisted breeding (GAB) mapping methods such as Marker Assisted Selection (MAS), Association mapping, Marker Assisted Backcross Breeding (MABC), Marker Assisted Recurrent Selection (MARS), Nested Association Mapping (NAM), Multi-parent Advanced Generation Intercross (MAGIC), and Genomic Selection (GS). The link between the different methods is that they use molecular markers to identify and select genes/QTLs linked to the trait of interest. The merits and demerits of the different methods are discussed below.

The MAS makes use of a biparental mapping population such as F_2, recombinant inbred lines (RIL) and near isogenic lines (NILs), for understanding the genetic architecture of a trait and to identify one or few major genes/quantitative trait loci (QTL). The linked markers for the traits are then used to introgress major genes or to pyramid a few target genes/ QTLs into elite varieties through backcrossing. There are several successful examples of MAS in chickpea (Varshney et al., 2013a,b; 2014; Ahmad et al., 2014), groundnut (Chu et al., 2011; Janila et al., 2016 a,b) and pearl millet (Hash et al., 2000; Hash and Witcombe, 2001). However, the application of MAS is restricted to those traits that are controlled by one or few major genes/ QTLs (Lee, 1995; Holland, 2004) and appears inefficient in dealing with polygenic traits which are controlled by several genes with minor effects and interactions of the genes with each other and the environment (Riedelsheimer et al., 2012). Also, the use of only two parents in population development restricts its ability to capture the allelic diversity and genetic background effects that are often present in many preferred traits because of different pedigrees. Validation of QTLs and closely linked markers in adapted germplasms as well as in populations developed using parents with different pedigrees is critical to the success of MAS.

Association mapping is another mapping technique that makes use of genotypic information on large and genetically diverse populations that are extensively phenotyped for target traits across locations and years (Jannink et al., 2001; Rafalski, 2002). The use of a breeding population reduces the need for extensive validation and allows QTL detection and allelic value estimation that can be directly used by MAS (Holland, 2004; Breseghello and Sorrells, 2006). However, the identification of QTLs and their effects can be largely influenced by low heritability of the trait, small population sizes, few large-effect QTLs, confounding population structures, and arbitrary significance thresholds (Beavis, 1998; Schon et al., 2004).

Marker assisted recurrent selection (MARS) was proposed as a suitable alternative to transfer QTL (usually minor) for polygenic traits and to enhance genetic gains (Lande and Thompson, 1990). It uses a two-step approach to enhance the efficiency of MAS: (i) select significant markers from large marker sets, and (ii) combine phenotypic information with significant markers in a selection index that explains a significant proportion of additive genetic variance. However, factors such as number and size of the families, population type and reliability of marker-trait associations can limit the effectiveness of the MARS approach (Mayor and Bernardo, 2009).

Multi-parent mapping populations such as nested association mapping (NAM) and multi-parent advanced generation intercross (MAGIC) populations provide a means of combining the advantages and eliminating the disadvantages of linkage analysis and association mapping. Both populations use a set of divergent parents to make crosses and develop populations which are then phenotyped in single/multiple environments. The use of an assorted set of parents increases the chances of recombination leading to a novel rearrangement of alleles and an increase in genetic diversity. MAGIC populations are an ideal resource for generating high-density maps using germplasms relevant to breeders as well as generating information on the complex architecture of many traits associated with crop performance and product quality (Cavanagh et al., 2008). However, a MAGIC population requires a greater use of resources and more time to develop. Also, the population is likely to exhibit extensive segregation for diverse traits. Another disadvantage is that a huge population size in the range of 1000s is very difficult to phenotype accurately in a non-biased manner.

In all the above approaches a significant set of markers is necessary to identify major as well as minor QTLs that are linked to the target trait and the information is used in making selection decisions. Genomic Selection (GS) also uses marker information and phenotypic data to develop prediction models and aids in making selection decisions. However, in GS as compared to other GAB methods (which mainly consider the markers linked to genes/QTLs of target traits of interest), whole genome marker data is used in a single step using a manageable population size (encompassing diversity for all economically important traits) to achieve maximally accurate marker effect estimates. The use of the whole genome approach ensures that all QTLs are in LD with at least one marker and combined with high-quality phenotypic data the method is useful in training prediction models and calculating the genomic estimated breeding values (GEBVs). The GS strategy does not require the need to map genes and search for linked QTL-marker loci associated individually. Depending on the prediction accuracy of the model, the GEBVs can be used to make selection decisions thereby reducing the need for phenotyping. Deployment of GS along with a rapid generation turnover helps in shortening the breeding cycle length and, improving selection intensities and accuracies as superior individuals are selected using marker data, thereby enhancing the genetic gain per unit of time and cost (Desta and Ortiz, 2014). Considering epistatic and genotype × environment interaction (GEI) in GS models enhances the precision of assessing the predictive ability that aids in the selection of lines across the environments for highly complex traits (Oakey et al., 2016).

Genomic selection in plant breeding

In GS, a population comprising of diverse genotypes differing for all the important target traits is genotyped using genome-wide markers and phenotyped for complex polygenic traits at single/multi-year or single/multi–locations depending on the objective of the breeder and resource availability. This population constitutes the training population (TP). The genotypic and phenotypic data on TP are used for designing a prediction model to be used in candidate/test populations using genotypic data only. The genetic resemblance between training and candidate/test populations ensures better prediction accuracies. As compared to other available alternatives, GS offers a promising approach for hard-to-phenotype complex traits in crops such as resistance/tolerance to biotic and abiotic stresses along with nutritional quality traits. Once the prediction models are established for each trait GEBVs can be used to predict performance thereby reducing the need for phenotyping. It also reduces the selection cycle length through limiting phenotypic selection to advance generations hence, reducing time and cost, leading to higher genetic gain.

To estimate the GEBVs and to develop GS models various parametric and nonparametric approaches among different statistical models have been explored (Meuwissen et al., 2001; de los Campos et al., 2009; 2010; Crossa et al., 2010; 2017; Jannink et al., 2010) and several studies were done to compare simulated and empirical data (Heslot et al., 2012; de los Campos et al., 2013; Howard et al., 2014). Simulation studies have revealed that the breeding value could be predicted with an accuracy of upto 0.85 from the marker data alone. Several reports have compared GS with other selection methods and found much better selection efficiency using GS than other genomics assisted breeding tools. For example, Bernardo and Yu (2007) reported 40% better selection efficiency in GS compared to MARS. Similarly, phenotypic selection (PS), MAS, and GS efficiencies across 13 different agronomic traits in winter wheat were compared and it was found that average prediction accuracy for GS was 28% higher as compared to MAS, and as accurate as PS (Heffner et al., 2011). Massman et al. (2013) found that GS was more advantageous over MARS and/or conventional selection in increasing genetic gains per unit of time (years) using a maize biparental with temperate and tropical maize in an empirical selection experiment. Use of GS resulted in 14–50% higher genetic gains than MARS for grain yield and stover quality traits. In another study involving eight biparental tropical maize populations evaluated in drought stressed environments reported superior GS over pedigree selection (Beyene et al., 2015). Arruda et al. (2016) compared efficiency of GS and MAS for fusarium head blight resistance in wheat and reported prediction accuracy ranged from 0.4–0.9 for GS whereas < 0.3 in MAS. Also, with the same selection intensity, GS resulted in higher selection differentials than MAS for all traits. All these studies indicate the efficiency of GS in regular breeding program over other methods. Genomic selection in crop species is more advantageous owing to its potential to predict selection decisions even during the off-season, hence enhancing significant genetic gain on a yearly basis (Heffner et al., 2010). Genomic selection has been successfully deployed in several crops, i.e., rice (Spindel et al., 2015), wheat (Rutkoski et al., 2012; Poland and Rife, 2012;

Battenfield et al., 2016), maize (Bernardo and Yu, 2007; Albrecht et al., 2011; Technow et al., 2013), soybean (Zhang et al., 2016).

Higher prediction accuracies are critical for the success of any GS breeding program. Multiple simulation and empirical studies involving estimation of prediction accuracies depends on multiple factors *viz.* marker type/number (Chen and Sullivan, 2003; Poland and Rife, 2012), population structure (Nakaya and Isobe, 2012; Spindel et al., 2015), size of training population (Daetwyler et al., 2008), type of traits and their heritability (Zhong et al., 2009; Zhang et al., 2016) and importantly the relationship between training populations and selection candidates. To achieve higher prediction accuracies and meeting diverse requirements, numerous GS prediction models have been proposed. Some of the commonly used models are: Random Regression Best Linear Unbiased Predictor (RR-BLUP; Meuwissen et al., 2001), Least Absolute Shrinkage and Selection Operator (LASSO) (Tibshirani, 1996), semi-parametric strategies (reproducing kernel Hilbert spaces) (Gianola and van Kaam, 2008), Bayesian approaches *viz.*, Bayesian LASSO (Park and Casella, 2008), Bayes A (Meuwissen et al., 2001), Bayes B (Meuwissen et al., 2001), Bayes Cp (Habier et al., 2011), machine learning based Random Forest Regression (RFR) (Breiman, 2001), and Support Vector Regression (SVR) (Drucker et al., 1997). Selection of an appropriate GS model differs from case to case; thus, multiple models should be tested to identify the most suited GS model for any crop/trait. The involvement of large GEI makes the traits more complex and limits the genetic gain. GS models which account for GEI effects in predictions of GEBVs help in making precise selections for even unobserved environments (Jonas and de Koning, 2013; Oakey et al., 2016; Roorkiwal et al., 2018).

For groundnut, chickpea and pearl millet, the adoption of GS in regular breeding program is just in the initial phases. Groundnut, chickpea and pearl millet are important crops of the semi-arid tropics. Until 2005, groundnut and chickpea were referred to as "orphan crops" due to availability of very limited genetic information (Varshney, 2016). Recently, there has been large-scale development of genomic resources in chickpea, groundnut and pearl millet including the availability of sequencing data. In case of chickpea, the whole genome sequence information of CDC Frontier, a kabuli chickpea variety was completed in 2013 (Varshney et al., 2013c) and in groundnut, sequence of two diploid progenitors A-genome (*Arachis duranensis*, accession V14167) and B-genome (*A. ipaensis*, accession K30076) and the cultivated groundnut are available (Bertioli et al., 2016; 2019; Chen et al., 2016; Zhuang et al., 2019). In pearl millet the draft genome sequence for the reference genotype Tift $23D_2B_1$-P_1 was made available (Varshney et al., 2017). The availability of huge marker data sets has increased the possibility of deploying GS in all the three crops and studies are in progress to develop GS prediction models in groundnut (Chaudhari et al., 2019), chickpea (Roorkiwal et al., 2016; 2018; Li et al., 2018) and pearl millet (Liang et al., 2018; Jarquin et al., 2020) (Table 1). This chapter elucidates on on-going efforts to deploy GS in groundnut, chickpea and pearl millet-three important ICRISAT mandate crops and discuss possibilities by integrating available genomic resources to harness the full potential of modern breeding approaches.

Groundnut

Groundnut or peanut (*Arachis hypogaea* L.), an important oilseed, food, fodder, and feed legume crop, plays an important role in securing the livelihood of smallholder and marginal farmers across the globe, especially in the semi-arid tropics. It is a self-pollinated allotetraploid ($2n = 4x = 40$), with an AABB genomic constitution, and is derived from a recent hybridization event involving the two diploid wild progenitors—*Arachis duranensis* and *Arachis ipaensis* (Kochert et al., 1996; Grabiele et al., 2012; Moretzsohn et al., 2013). Groundnut is cultivated across 114 countries over an area of 28.51 m ha with a total production of 45.95 m t and an average yield of 1611 kg/ha (FAOSTAT, 2014–18). The crop has achieved around 198% (29.80 m t) increase in global groundnut production largely due to the combined effect of an increase in area by 56.3% (9.78 m ha) and yield by 91.0% (787.2 kg/ha) during last five and a half decades (Nigam et al., 2021). Besides, its global significance in subsistence farming and research efforts, the genetic gains for yield in many regions are still low with average growth rates of 0.5% in Africa and 1.2% in South Asia as compared to America (1.6%) and East-Asia (2.7%). The productivity levels of major groundnut growing countries of Africa (Nigeria, Sudan, Tanzania, Senegal, Chad, and Niger) and Asia (India, Myanmar and Bangladesh) are low as compared to the developed nations. In these countries, the crop is mainly grown under a rainfed ecosystem with limited resources, inputs and age-old varieties which leads to reduced productivity (Nigam et al., 2021). These productivity issues are further challenged by their levels of susceptibility to important biotic and abiotic stresses.

Among the biotic stresses, foliar fungal diseases, i.e., late leaf spot (LLS) caused by *Phaeoisariopsis personata* (Berk. & Curt.) Van Arx, early leaf spot (ELS) caused by *Cercospora arachidicola* Hori and rust caused by *Puccinia arachidis* Spegazzini are the major production constraints (Janila et al., 2013a). In addition to these, stem and pod rot caused by *Sclerotium rolfsii*, is an emerging potential threat in many warm, humid areas, especially where irrigated cultivation is expanding. Bacterial wilt caused by *Ralstonia solanacearum* is predominant among bacterial diseases in South-East Asia, particularly China and Vietnam (Liao, 2014; Zhao et al., 2016). The intensities

Table 1. Deployment of GS in groundnut, chickpea and pearl millet.

Crop	Training population size	Models employed	Selected traits for GS	Prediction accuracy	References
Groundnut	336 diverse groundnut lines	Bayesian Generalized Linear Regression (BGLR)	Foliar disease resistance, oil, protein and oleic acid content, and yield traits	0.40 to 0.60 (using different models)	Pandey et al. (2020a)
Chickpea	320 elite breeding lines	RR-BLUP, Kinship GAUSS, Bayes Cπ, Bayes B, Bayesian LASSO, Random Forest	Yield related traits- Seed yield, 100 seed weight, days to 50% flowering, days to maturity	0.138 to 0.912 (using different models)	Roorkiwal et al. (2016)
	132 varieties and advanced breeding lines	RR-BLUP, Bayesian LASSO, Bayesian ridge regression	Grain yield, 100 seed weight, empty pod ratio, podding time score, flowering emergence score, early vigor score under drought stress	0.25 to 0.8 (using different subsets of SNPs based on p-values from GWAS results)	Li et al. (2018)
	320 elite breeding lines	Multiplicative reaction norm model (MRNM)	100 seed weight, biomass, days to 50% flowering, days to maturity, harvest index, plant height, number of plant stand, seed yield	0.053 to 0.773 (using different models and cross-validation scheme 2 for 8 traits	Roorkiwal et al. (2018)
Pearl millet	320 hybrids and 37 inbred parents	Best linear unbiased prediction (BLUP)	Days to 50% flowering, plant height, grain yield and 1000 grain weight	0.73–0.74 (1000-grain weight), 0.87–0.89 (days to flowering time), 0.48–0.51 (grain yield) and 0.72–0.73 (plant height)	Liang et al. (2018)
		M1-General Combining Ability (GCA) model M2-General plus Specific Combining Ability Model M3- General plus Specific Combining Ability in Interaction with Environments Model	Grain Yield		Jarquin et al. (2020)
	994 genotypes (963 inbred lines and 31 wild accessions)	Ridge Regression BLUP (RR-BLUP)	Grain yield for test crosses, Hybrid performance	0.6 for grain yield; 170 promising hybrids detected	Varshney et al. (2017)

of different biotic stresses often depend on unfavorable abiotic environments. For example, terminal drought also helps in predisposing the produce to *Aspergillus flavus* infection in the field leading to aflatoxins contamination which are potent carcinogens thus affecting the quality of produce and contributing to significant health risk (Janila et al., 2013b). Among the abiotic stresses, drought and heat stress can adversely affect the yield across semi-arid ecologies (Prasad et al., 2010). Hence, the development of climate-smart cultivars that offer tolerance to drought and heat stresses has become an important objective for groundnut breeding programs.

Genetics of rust resistance revealed recessive digenic inheritance (Vindhiyavarman et al., 1993), dominant single-gene resistance (Singh et al., 1984) whereas; LLS resistance is complex and governed by polygenes (Dwivedi et al., 2002) with a combination of both nuclear and maternal gene effects (Janila et al., 2013a). Mapping studies carried out on a recombinant inbred line (RIL) population identified a major QTL contributing 67% and 80% phenotypic variation to LLS and rust respectively (Khedikar et al., 2010; Sujay et al., 2012; Kolekar et al., 2016). These QTLs were introgressed into elite varieties and reported moderate levels of resistance which could be attributed to the absence of minor effect QTLs/genes (Janila et al., 2016a). Furthermore, the high G × E interactions and environment effect make these traits more complex in nature. The complex nature of resistance to these diseases makes the identification of resistant and susceptible lines cumbersome through conventional screening techniques (Leal-Bertioli et al., 2009). Hence, for achieving higher genetic gains for resistance to LLS and rust, both major and minor QTL/gene effects need to be captured which is possible through GS. Similarly, the drought, heat and other abiotic stresses along with yield and nutritional quality traits are polygenic traits whose expressions are influenced by their interactions with the growing environment (Krishnamurthy et al., 2007; Janila et al., 2013b). For polygenic traits, crop improvement through conventional breeding strategies is difficult to achieve as it requires a large population size along with expenditure on time and resources for phenotyping. In contrast genomics-based approaches help in improving selection efficiency and accelerating the rate of varietal development (Pandey et al., 2012; Varshney et al., 2013a). Several genomic resources developed and deployed in groundnut during past decades resulted in enhancing precision in selections and reduced duration of variety development (Pandey et al., 2016; Varshney et al., 2019; Pandey et al., 2020b).

There are three genomic assisted crop improvement approaches used by breeders in making selection decisions, i.e., MAS, MARS and GS. The MAS makes use of closely linked markers to identify and select plants that carry the target gene/QTL and this is already being deployed in groundnut to improve nematode resistance (Chu et al., 2011), resistance to rust and LLS (Kolekar et al., 2017; Janila et al., 2016b), and oleic acid content (Janila et al., 2016a). MAS and MARS which are based on biparental populations to identify trait linked markers and robust validation are often limited by diagnostic marker availability for complex target traits. In MAS a few diagnostic markers are used to follow the inheritance of specific loci influencing a trait whereas in MARS a subset of significant markers is used to select for QTL in a population. In contrast, GS uses multiple genetic markers linked to all the loci that are influencing the trait (Oakey et al., 2016). Hence, GS is becoming the most preferred approach for deployment in groundnut improvement due to its higher selection intensity and improved accuracy through capturing the effect of large as well as small-effect QTLs thus leading to better genetic gains in breeding programs for complex polygenic traits (Meuwissen et al., 2001; Heffner et al., 2010; Bernardo, 2010; Shikha et al., 2017; Wang et al., 2019). The approach of GS is relatively new to groundnut as compared to other model species. TPs have been developed which are being phenotyped in multiple environments to develop accurate prediction models by involving GEI components to improve upon the selection efficiency. The case study presented below reports on the development of TP, phenotyping and genotyping strategies employed and establishment of prediction models in groundnut. The strategy employed is depicted in Fig. 1.

Developing GS training population in groundnut

The TP must represent/resemble the breeding population with a maximum proportion of trait variance associated with the markers along with low collinearity between markers. Considering the above-mentioned criteria, a genomic selection training population (GSTP) comprising of 340 genotypes was selected based on the diversity available for morphological and important economic traits in different subspecies and botanical varieties of cultivated groundnut and constituted at the International Crops Research Institute for the Semi-Arid Tropics (ICRISAT), Hyderabad India. The GSTP includes elite breeding lines and germplasms from the groundnut breeding program of ICRISAT, and University of Agricultural Sciences (UAS) in Dharwad, Indian Council of Agricultural Research-Directorate of Groundnut Research (ICAR-DGR) along with accessions from the gene bank of ICRISAT and popular cultivars from India. The population includes genotypes representing both the subspecies, i.e., *A. hypogaea* ssp. *fastigiata* (227) and *A. hypogaea* ssp. *hypogaea* (113) representing 21 geographically diverse countries. The genotypes of *A. hypogaea* ssp. *fastigiata* include representations from all the four botanical varieties *A. hypogaea* ssp. *fastigiata* var. *vulgaris* (212), *A. hypogaea* ssp. *fastigiata* var. *fastigiata* (10), *A. hypogaea* ssp. *fastigiata* var. *peruviana* (4) and *A. hypogaea* ssp. *fastigiata* var. *aequatoriana* (1)

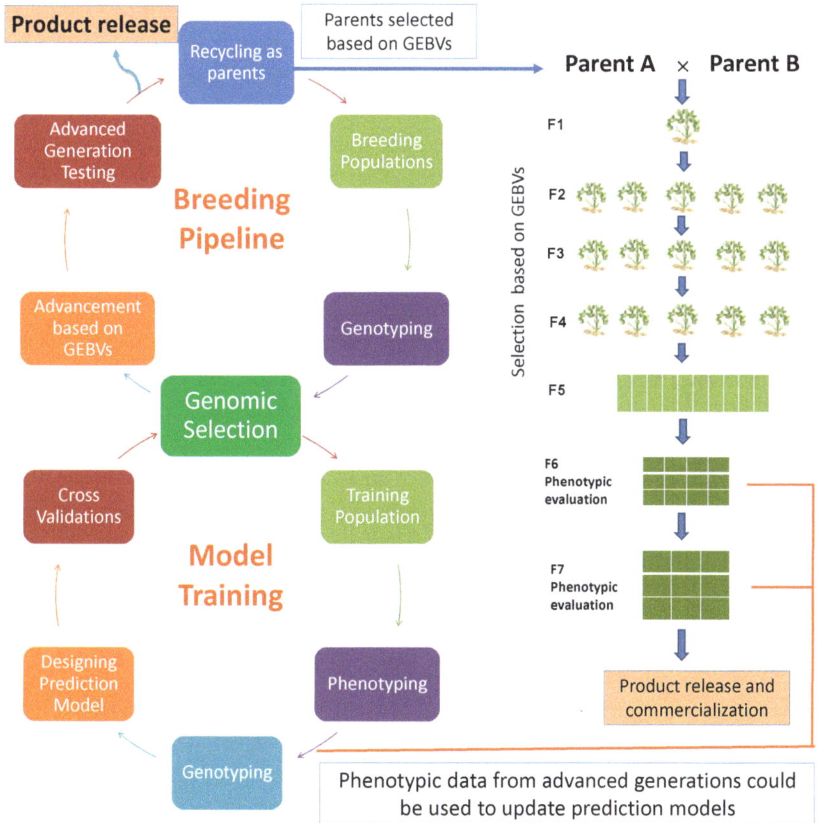

Fig. 1. Proposed workflow of development and deployment of genomic selection in groundnut. In the model training cycle, a training population is phenotyped and genotyped for designing a genomic selection prediction model which is set for use in the breeding pipeline after cross validations for prediction accuracies. The breeding pipeline uses genotyping data on section candidates for GS to make advancement decisions based on GEBVs. The lines with higher GEBVs are forwarded for advance generation testing across target environments for product release and/or to be recycled into the breeding program. The breeding scheme on the right showing deployment of GS along with rapid generation advancement could significantly shorten the breeding cycle length in groundnut.

whereas the genotypes of subspecies *A. hypogaea* ssp. *hypogaea* represent botanical varieties *A. hypogaea* ssp. *hypogaea* var. *hypogaea* (111) and *A. hypogaea* ssp. *hypogaea* var. *hirsuta* (1) (Chaudhari et al., 2019).

Case studies on deployment of GS in groundnut

A training population of 340 lines was phenotyped for 11 agronomic, 7 quality and 6 foliar disease resistance traits at Patancheru, Aliyarnagar and Jalgaon, India during 2015–16. The genotyping of 318 genotypes using a high density '*Axiom_Arachis*' array containing > 58,000 highly informative genome-wide single nucleotide polymorphism (SNP) markers identified 13,355 polymorphic SNPs. The phenotypic and genotypic data were analyzed with four different GS models which were tested with three cross-validations (CV) (Pandey et al., 2020a). Of these four models, M1 (Model 1) (E+L) includes the main effects of environments (E) and lines (L); M2 (Model 2) (E+L+G) includes the main effects of markers (G) in addition to environments (E) and lines (L); M3 (Model 3) (E+L+G+GE) referred to as a naïve interaction model includes GEI additionally; and M4 (Model 4) (E+L+G+LE+GE), a naïve and informed interaction model offers an advantage of estimating L × E interactions over M3.

The analysis was performed with the Bayesian Generalized Linear Regression (BGLR) R-package using four GS models (de los Campos et al., 2013; Perez-Rodriguez et al., 2015). The results highlight better prediction accuracies estimated from models M2, M3 and M4 as compared to model M1 indicating the advantage of using markers. Better prediction accuracies (> 0.600) were reported for agronomic traits, i.e., days to 50% flowering, hundred seed weight and days to maturity; nutritional quality traits, i.e., oil content, oleic and linoleic acid content, and resistance to rust and late leaf spot at 90 and 105 days after sowing (Pandey et al., 2020). The comparative prediction accuracy for different GS models to perform selection for untested genotypes, and unobserved/unevaluated environments could provide greater insights on potential application of GS in groundnut breeding.

Another GS report on groundnut used 662 recombinant inbred lines (RIL) from three crosses Florida 07 × GP NC WS 16 (383 RILs), Florida 07 × C76-16 C1798 (127 RILs) and Tifrunner × C76-16 (152 RILs). Genotyping using SNPs and SSR markers along with phenotyping data of eight traits on a subset of these populations were used for GS predictions. The prediction accuracies were validated comparing the estimated GEBVs with the actual trait values. They have reported prediction accuracies ranging from 0.21 for tomato spotted wilt virus (TSWV) to 0.79 for LLS indicating the robustness of the models for GS (Gimode et al., Unpublished).

Efforts were also made to construct GS models for aflatoxin resistance using 625 RILs from two populations that were genotyped using the Affymetrix v2 SNP chip and phenotyped for aflatoxin resistance over three years. The results showed low prediction accuracies (0.2 to 0.3) due to the low heritability values (0.2 and 0.1, respectively) in both the populations (Gimode et al., Unpublished).

Prospects of GS in groundnut

Breeding approaches and genomic tools to develop superior crop varieties have been evolving over time for achieving higher genetic gains for target traits. Besides the success stories made through conventional breeding approaches, genomic-assisted breeding methods such as MAS or MABC have been successfully used to develop groundnut varieties with improved oleic acid content and resistance to rust and late leaf spot (Janila et al., 2016a,b; Pandey et al., 2016; Varshney et al., 2019). The recent achievements in genome sequencing of diploid progenitors (Bertioli et al., 2016; Chen et al., 2016) and the subspecies (*A. hypogaea* ssp. *hypogaea* and *A. hypogaea* ssp. *fastigiata*) of cultivated tetraploid groundnuts (Bertioli et al., 2019; Chen et al., 2019; Zhuang et al., 2019) led to the development of high and mid-density genotyping assay (Pandey et al., 2017; Clevenger et al., 2017) required to use GS in the breeding program. Phenotypic selection for complex traits in groundnut such as yield, resistance to foliar fungal diseases, and tolerance to drought and heat stresses requires huge resources in maintaining a large population size in early generations and conducting replicated yield trials in advanced generations across target environments due to the bulky nature of the crop. Including available as well as new breeding lines of groundnut for this diverse set of traits into the TP will be useful in developing GS models with higher prediction accuracy. These breeding lines could also be used as parents in a crossing program based on their GEBVs.

Besides biotic and abiotic stresses, agronomic traits such as days to maturity, number of pods per plant, shelling outturn, hundred seed weight and yield along with nutritional quality traits such as oil and protein content are the key priority traits in groundnut governed by polygenes and are complex in nature. Genomic selection in future would serve as an efficient selection tool to improve these complex traits in groundnut.

Chickpea

Chickpea (*Cicer arietinum*) is the second largest cultivated grain legume globally after soybean. It is considered a protein rich crop due to its high protein content of around 17–30% by dry weight and serves as a vital source of protein in vegetarian diets. Being a legume, chickpea has the ability to fix atmospheric nitrogen and improve upon the soil nutritional profile through crop rotation. It is a self-pollinated, diploid ($2n = 16$) annual crop with genome size of ~ 740 Mbp (Varshney et al., 2013b). Currently, the chickpea is grown over 14.6 million ha area across 55 countries globally with an annual production of 14.8 million tons (FAOSTAT, 2018). Being cultivated under barren and low productive soils in the arid and semi-arid regions, the crop is exposed to several biotic and abiotic stresses. Terminal drought and heat are the major abiotic stresses whereas Fusarium wilt, Ascochyta blight and pod borer are the major biotic stresses hampering the chickpea production globally.

Commercially, two distinct forms of the cultivated chickpea are available: desi chickpeas are small, angular shaped and have colored seeds containing a high fiber percentage whereas the kabuli types are large, owl shaped, beige-colored seeds with low fiber percentage. In terms of their geographic distribution-desi types are found in central Asia and in the Indian subcontinent while the kabuli types occur in the Mediterranean region.

The sequencing of the kabuli chickpea variety "CDC Frontier" in 2013 (Varshney et al., 2013b) has resulted in the development of large-scale genome-wide genetic markers and the use of high-throughput genotyping platforms to obtain precise, rapid and cost-effective genotyping solutions. For instance, informative SNPs with high density are being chosen and used to design assays/platforms in chickpea (Hiremath et al., 2012; Roorkiwal et al., 2013). The development and deployment of different genotyping platforms such as Genotyping by Sequencing (GBS) (Deokar et al., 2014; Jaganathan et al., 2015; Verma et al., 2015), DArT assay (Thudi et al., 2011) provide cost effective and precise genotyping solutions to chickpea (Varshney et al., 2017).

Deployment of GS in chickpea breeding

GS was deployed in developing superior lines with higher yields under different environmental conditions and enhancing the genetic gains in chickpea breeding programs. The availability of draft genome sequences (Jain et al., 2013; Varshney et al., 2013b; Ruperao et al., 2014; Gupta et al., 2016) along with whole genome re-sequencing data for several hundred lines in chickpea has resulted in a manifold increase in marker densities which are being used to address selection issues for complex traits through GS. The first published report on the application of GS in chickpea by Roorkiwal et al. (2016) discussed the development of suitable GS models for predicting breeding values using genome wide markers on elite breeding lines in chickpea. The TP developed at ICRISAT, Patancheru comprised a set of 320 elite chickpea breeding lines (including both desi and kabuli types) from the International Chickpea Screening Nursery (ICSN). Phenotyping of the population was done for yield and yield related traits at two locations in India, i.e., IARI, New Delhi and ICRISAT, Patancheru for two seasons. Genotyping of the population was done under the DArTseq platform using 3000 polymorphic markers. Six different statistical models including Ridge Regression Best Linear Unbiased Predictor (RR-BLUP), Kinship GAUSS (semiparametric model), Bayes Cπ, Bayes B, Bayesian Least Absolute Shrinkage and Selection Operator (Bayesian LASSO) and Random Forest (RF) (machine learning algorithm) were used for prediction of GEBVs for four yield related traits *viz*. seed yield, 100 seed weight, days to 50% flowering and days to maturity. The prediction accuracies using these models varied from 0.138 (seed yield) to 0.912 (100 seed weight) (Roorkiwal et al., 2016).

In another study carried out by Li et al. (2018) on chickpea, it was reported that incorporating the results of Genome Wide Association Studies (GWAS) into the prediction model can significantly increase prediction accuracy with a small population size. The TP used in the study comprised of 13 Australian released varieties and 119 Australian and Indian-derived breeding lines, selected for yield potential and adaptation to drought prone environments. The technique of GWAS was used to identify candidate genes/SNPs significantly associated with yield and yield-related traits under drought stress. Significant association was found between SNPs and yield and yield related traits under drought-prone environments. Four genetic regions were identified as containing SNPs significantly associated with multiple traits which was indicative of pleiotropic effects. The results from GWAS were incorporated into the GS model to develop prediction equations. Three different models: RR-BLUP, Bayesian least absolute shrinkage (Bayesian LASSO or BL) and Bayesian ridge regression (BRR) were used to calculate GEBVs. Based on *p*-values from GWAS results the SNPs were categorized into different subsets. Prediction accuracies for four traits-grain yield per ha (GY), hundred seed weight (HSW), seed number per plant (SN) and early vigor (EV) score increased by more than twofold (prediction accuracy > 0.5 for SN to 0.8 for HSW) when RR-BLUP was combined with subsets of SNPs with *p*-values between 0.05 and 0.01. The subset contained SNPs that were significantly associated with the traits. When using subsets of SNPs with *p*-values of 3.45e-07 (equal to 0.05 after Bonferroni correction) the prediction accuracy decreased rapidly. However, the lowest prediction accuracy (≥ 0.25) was observed for GY, HSW and EV when all SNPs were used to make predictions.

GS studies based on single location/single season phenotypic data can lead to biased selection of stable lines especially for those traits which are having significant GEIs. Inclusion of GEI in genomic prediction models can improve upon the selection efficiency as it is possible to select lines with optimal overall performance across environments and in specific target environments as well. However, the use of single location/season data or multi-environment data to develop prediction models relies mainly on the objective of the breeder. GS models that consider GEIs were proposed by Burgueno et al. (2012), Jarquin et al. (2014) and Oakey et al. (2016) and are being used for chickpea breeding to improve upon the prediction accuracy (Roorkiwal et al., 2018). By using three different cross-validation (CV) schemes to mimic field scenarios faced by a breeder it was found that a cross-validation scheme that includes lines which have been evaluated in some of the environments but not in other environments gave maximum prediction accuracy for different traits and different models. DArTseq was found to be more accurate than GBS and the combined genotyping set (DArTSeq and GBS) regardless of the cross-validation scheme with most of the main effect marker and interaction models (Roorkiwal et al., 2018).

Pearl millet

Pearl millet [*Cenchrus americanus* (L.) Morrone; Syn. *Pennisetum glaucum* (L.) R. Br.], grown for food, feed and fodder, is cultivated on about 32 million ha in more than 30 countries of four continents, namely, Asia, Africa, North America, and Australia (Yadav et al., 2016). It is a diploid (2n = 2x = 14), cross-pollinated warm-season crop with a genome size of 1.76 Gb with tremendous photosynthetic potential and high biomass production capacity (Varshney et al., 2017). Pearl millet is majorly grown in the arid and semi-arid tropical regions of Africa (14 million ha) and Asia (12 million ha) (Yadav et al., 2016). In Africa, the majority of pearl millet acreage is in Western and Central Africa where it is grown in 17 countries. Niger, Nigeria, Burkina Faso, Mali and Senegal account for nearly 90% of cultivated pearl millet area in Africa. At an individual country level, India is the largest producer of this crop with about 9 million ha area and 8.5 million tons of grain production. The major pearl millet growing Indian states of Rajasthan, Maharashtra, Gujarat, Uttar Pradesh and Haryana, account for > 90% acreage of pearl millet (Yadav et al., 2016).

Pearl millet, a C_4 plant belonging to the family Poaceae, has high photosynthetic efficiency and dry matter production capacity. It is predominantly a cross-pollinated crop with 75–80% outcrossing (Burton, 1983). Pearl millet is better adapted to growing areas characterized by dry semi-arid environments than other cereals. It has the highest water use efficiency under water–limiting environments (Zegada-Lizarazu and Iijima, 2005), highest level of tolerance to high-temperatures during the reproductive phase (Gupta et al., 2015) and high salinity tolerance (Kulkarni et al., 2006; Yadav et al., 2012 a,b). The characteristics like high genetic variability, protogyny and availability of efficient cytoplasmic genetic male sterility system offer great possibilities to exploit heterosis for both grain and fodder yield through hybrid development (Andrews and Kumar, 1992; Rai et al., 1999; Rai et al., 2001; Yadav et al., 2012a).

Pearl millet is a nutritious and well digested source of calories and proteins for humans. It contains about 12–14% protein, 5% fat, 67% carbohydrates and 360 K cal energy value 100^{-1} g of grains (Ejeta et al., 1987; Khairwal et al., 1999). The amino acid profile of pearl millet grain is more balanced than that of normal sorghum or maize, and is comparable to those of wheat, barley and rice. The lysine content of the protein ranges from 1.9 to 3.9 g/100 g protein

(Ejeta et al., 1987). The energy value of pearl millet grain is relatively high, arising from its higher oil content relative to maize, wheat or sorghum (Hill and Hanna, 1990). Collins et al. (1997) reported that commercial layers of the given feed containing pearl millet grain had lower omega-6 to omega-3 fatty acid ratio, endowing the eggs with a fatty acid profile more favorable to human health. Pearl millet is also an excellent forage crop given its lower hydrocyanic acid content than sorghum. Its green fodder is rich in protein, calcium, phosphorus and other minerals with oxalic acid within safe limits (Yadav et al., 2012a).

Recent studies have shown the presence of a wide range of genetic variability for grain iron (Fe) content (18 to 135 mg/kg) and zinc (Zn) content (22 to 92 mg/kg) (Velu et al., 2008a; 2008b; Rai et al., 2012; Govindaraj et al., 2013; Kanatti et al., 2014), indicating good prospects for genetic enhancement of Fe and Zn. For example, an improved and high-Fe biofortified version of the pearl millet variety ICTP 8203 was released as Dhanashakti (ICTP 8203 Fe 10-2) in 2014 for cultivation in India. This variety had 71 mg/kg of Fe density (9% more than ICTP 8203) and 2.2 t/ha grain yield (11% more than ICTP 8203), with no change in Zn density (38–39 mg/kg) and flowering time (45 days) (Rai et al., 2014).

The availability of the draft genome sequence of pearl millet for the reference genotype Tift $23D_2B_1$-P_1 (Varshney et al., 2017) along with resequencing of 994 genotypes involving 963 inbred lines and 31 wild accessions has made it possible to begin deploying GS to predict both the general combining ability (GCA) of new inbred parents and the specific performance of particular pearl millet hybrids (Liang et al., 2018). Resequencing data in particular was used to establish marker trait associations, to predict hybrid performance and to define heterotic pools.

Deployment of genomic selection in pearl millet

In pearl millet improvement, breeders face multiple challenges in efficiently breeding new elite parental lines based on inbred phenotypic data, balancing trade-offs between optimal forage yield and quantity with optimal grain yield and quality, and significant GEIs for complex traits resulting in inconsistent response patterns (Jarquin et al., 2020). Use of GS tools in pearl millet breeding programs has the potential to address all these challenges along with facilitating precise prediction of hybrid performance and ideal resource allocation (Jarquin et al., 2020; Srivastava et al., 2020). The application of GS strategies in pearl millet breeding is discussed based on available literature.

Varshney et al. (2017) used the resequencing data obtained from 994 pearl millet lines for carrying out GS to predict grain yield for test crosses. Prediction scenarios involving the performance of grain yield under control, early stress, late stress and across environments were investigated. A high prediction accuracy of 0.6 was obtained for the performance of grain yield across environments. Prediction accuracy was measured as a Pearson correlation coefficient between the predicted and observed values, standardized with the square root of the heritability. At this level of prediction accuracy modelling studies have indicated that GS could substantially improve selection gain per year (Longin et al., 2015). Varshney et al. (2017) also followed a GS strategy that considers additive and dominance effects to predict hybrid performance in pearl millet. The RR-BLUP method was trained for prediction using the grain yield data obtained from replicated trials of 64 pearl millet hybrids grown in five environments in India during the time period 2004–2013. A set of 302,110 single nucleotide polymorphic markers (SNPs) identified in 580 maintainer (B-) and restorer (R-) lines of pearl millet was used to analyze the grain yield data and predict hybrid performance. The study identified 170 promising hybrid combinations of which 11 combinations were already being used and the remaining 159 are potential sources for hybrid production. Further analysis of predicted hybrid performance by including all possible 167,910 single-cross combinations revealed two sets of lines that are predicted to have an average 8% higher hybrid performance when crossed to each other and suggested that these hybrid combinations could serve as a nucleus in establishing heterotic pools for hybrid pearl millet breeding (Varshney et al., 2017).

Genotyping strategies and selection of appropriate prediction schemes are a crucial aspect of GS because of their influence on prediction accuracies. Liang et al. (2018) evaluated the potential of restriction-site associated DNA (RAD-seq) (Miller et al., 2007) and tunable GBS (tGBS) (Ott et al., 2017) genotyping strategies by combining them with four GS schemes in pearl millet. The phenotyping data for this study was collected from 320 hybrids and 37 inbreds which were evaluated in replicated trials at four locations in India in the year 2015. Phenotypic data was collected for four traits *viz.* days to 50% flowering (days), plant height (cm), 1000 seed weight (g) and grain yield (kg ha^{-1}). The RR-BLUP model was trained using genotyping and phenotyping data and used to estimate prediction accuracies in each of the four schemes. Scheme 1 (M1) involved, total set of genotyped and phenotyped inbreds which were divided into 5 equal groups. Scheme 2 (M2) followed conventional five fold cross validation in which the total set of genotyped and phenotyped hybrids was divided into 5 equal groups. The other two schemes followed the same system as M2, with the addition of all genotyped and phenotyped inbreds to the training dataset for all five subpredictions which either used BLUPs calculated across all individuals (M3A) or BLUPs calculated separately for inbred and hybrid populations (M3B). When data from only hybrids (M2) were used in the prediction model, the prediction accuracies (median) ranged from 0.73–0.74 (1000-grain weight), 0.87–0.89

(days to flowering time), 0.48–0.51 (grain yield) and 0.72–0.73 (plant height). For traits with little or no heterosis, the scheme involving phenotyping data from hybrid only and hybrid/inbred prediction schemes (M2 and M3A) performed almost equivalently.

Another recent study by Jarquin et al. (2020) implemented three GS models and also evaluated these models for grain yield using three different CV schemes mimicking real life situations observed in breeding programs: CV0 predicts the performance of hybrids in unobserved environments, CV1 predicts the performance of untested hybrids and CV2 resembles an incomplete field trial. They used genotypic (C) [conventional GBS RAD-seq] and T (Tunable GBS tGBS) and phenotypic data for grain yield obtained from Liang et al. (2018) for their study. Three genomic prediction models used were M1 (GCA Model – $E + G_{P1} + G_{P2}$), M2 (General and specific combining ability model – $E + G_{P1} + G_{P2} + G_{P1 \times P2}$) and M3 (General and specific combining ability in interaction with environments model – $E + G_{P1} + G_{P2} + G_{P1 \times P2} + G_{P1} \times E + G_{P2} \times E + G_{P1 \times P2} \times E$). Based on these models and cross validation scheme results, overall predictive ability was better for T platform than C platform, especially with the most comprehensive model M3. For predicting performance of new genotypes in the tested environments (CV1) as well as for prediction of test hybrids in unobserved environments (CV0), the T platform was consistent and best results were obtained with M3.

Prospects of GS in pearl millet

Pearl millet is an important climate smart nutritious cereal crop which remained neglected until the turn of the 21st century. Since then the rapid advancements in deploying genetic and genomic tools for pearl millet breeding has made it possible for breeders worldwide to use these molecular tools in their regular breeding program. Particularly, the advent of sequencing based genotyping tools such as genotyping-by-sequencing (GBS), RAD-sequencing, tGBS, and whole-genome re-sequencing (WGRS) has resulted in the generation of millions of genome-wide SNPs (Srivastava et al., 2020). The pearl millet inbred germplasm association panel (PMiGAP) developed at ICRISAT in partnership with Aberystwyth University contains germplasm lines, landraces and breeding lines from 27 countries and is considered a rich resource for mapping complex polygenic traits in pearl millet such as drought tolerance, grain Fe and Zn content, nitrogen use efficiency, components of endosperm starch, grain yield among others (Sehgal et al., 2015). With the availability of such diverse phenotypic and genotypic resources there is a renewed interest among the pearl millet breeders and genomic scientists to validate the loci linked to various economically important target traits and move from phenomics assisted breeding to genomics assisted breeding especially for complex polygenic traits.

Conclusion

Genomic selection (GS) which was initially deployed in animal breeding has since been extended to plant breeding for enhancing the rate of genetic gain by reducing long duration selection cycles and increasing the selection intensity and accuracy. GS can play a significant role in improving the traits with complex mechanisms involving a large number of small effect QTLs and their interactions with the environment. Preliminary work in GS for groundnut, chickpea and pearl millet improvement has produced encouraging results. In the case of groundnut a high density '*Axiom_Arachis*' array containing > 58K highly informative genome-wide SNP markers was found useful for genotyping the training population, for chickpea DArT marker systems gave more encouraging results while for pearl millet tunable genotyping by sequencing (tGBS) was recommended. Development and deployment of models that take into account epistatic interactions along with genotype by environment interactions could help to improve upon the prediction accuracies and aid in selection programs. Emerging technologies such as rapid generation advancement (RGA)/speed breeding along with incorporation of new genotyping tools and better statistical models will be required in future to realize the full potential of utilizing GS for developing improved varieties at a faster pace.

References

Ahmad, Z., Mumtaz, A. S., Ghafoor, A., Ali, A., Nisar, M. et al. 2014. Marker Assisted Selection (MAS) for chickpea *Fusarium oxysporum* wilt resistant genotypes using PCR based molecular markers. *Mol. Biol. Rep.* 41: 6755–6762. https://doi.org/10.1007/s11033-014-3561-3.

Albrecht, T., Wimmer, V., Auinger, H., Erbe, M., Knaak, C. et al. 2011. Genome-based prediction of testcross values in maize. *Theor. Appl. Genet.* 123: 339. https://doi.org/10.1007/s00122-011-1587-7.

Andrews, D. J. and Kumar, K. A. 1992. Pearl millet for food, feed and forage. *Adv. Agron.* 48: 89–139.

Arruda, M. P., Lipka, A. E., Brown, P. J., Krill, A. M., Thurber, C. et al. 2016. Comparing genomic selection and marker-assisted selection for Fusarium head blight resistance in wheat (*Triticum aestivum* L.). *Mol. Breeding* 36: 84.

Battenfield, S. D., Guzmán, C., Gaynor, R. C., Singh, R. P., Peña, R. J. et al. 2016. Genomic selection for processing and end-use quality traits in the CIMMYT spring bread wheat breeding program. *The Plant Genome* 9(2). https://doi.org/10.3835/plantgenome2016.01.0005.

Beavis, W. D. 1998. QTL analyses: Power, precision, and accuracy. pp. 145–162. *In*: Patterson, A. H. [ed.]. *Molecular Dissection of Complex Traits*. CRC Press, Boca Raton, FL.

Bernardo, R. and Yu, J. 2007. Prospects for genome wide selection for quantitative traits in maize. *Crop Sci*. 47: 1082–1090. https://doi.org/10.2135/cropsci2006.11.0690.

Bertioli, D. J., Cannon, S. B., Froenicke, L., Huang, G., Farmer, A. D. et al. 2016. The genome sequences of *Arachis duranensis* and *Arachis ipaensis*, the diploid ancestors of cultivated peanut. *Nat. Genet*. 47: 438.

Bertioli, D. J., Jenkins, J., Clevenger, J., Dudchenko, O., Gao, D. et al. 2019. The genome sequence of segmental allotetraploid peanut *Arachis hypogaea*. *Nat. Genet.* 51: 877–884.

Beyene, Y., Semagn, K., Mugo, S., Tarekegne, A., Babu, R. et al. 2015. Genetic gains in grain yield through genomic selection in eight bi-parental maize populations under drought stress. *Crop Sci.* 55: 154–163.

Breiman, L. 2001. Random forests. *Machine Learning* 45: 5–32.

Breseghello, F. and Sorrells, M. E. 2006. Association mapping of kernel size and milling quality in wheat (*Triticum aestivum* L.) cultivars. *Genetics* 172: 1165–1177.

Burgueño, J., de los Campos, G., Weigel, K. and Crossa, J. 2012. Genomic prediction of breeding values when modeling genotype × environment interaction using pedigree and dense molecular markers. *Crop Sci.* 52: 707–719.

Burton, G. W. 1983. Breeding pearl millet. *Plant Breed Review* 1: 162–182.

Cavanagh, C., Morell, M., Mackay, I. and Powell, W. 2008. From mutations to MAGIC: resources for gene discovery, validation and delivery in crop plants. *Curr. Opin. Plant Biol.* 11: 215–221.

Chaudhari, S., Khare, D., Patil, S. C., Sundravadana, S., Variath, M. T. et al. 2019. Genotype × environment studies on resistance to late leaf spot and rust in genomic selection training population of peanut (*Arachis hypogaea* L.). *Front. Plant Sci.* 10: 1338.

Chen, X. and Sullivan, P. F. 2003. Single nucleotide polymorphism genotyping: biochemistry, protocol, cost and throughput. *Pharmacogenomics J.* 3: 77–96.

Chen, X., Li, H., Pandey, M. K., Yang, Q., Wang, X. et al. 2016. Draft genome of the peanut A-genome progenitor (*Arachis duranensis*) provides insights into geocarpy, oil biosynthesis and allergens. *PNAS* 113: 6785–6790.

Chen, X., Lu, Q., Liu, H., Zhang, J., Hong, Y., Lan, H., Li, H., Wang, J., Liu, H., Li, S. and Pandey, M. K. 2019. Sequencing of cultivated peanut, *Arachis hypogaea*, yields insights into genome evolution and oil improvement. *Molecular Plant* 12(7): 920–934.

Chu, Y., Wu, C. L., Holbrook, C. C., Tillman, B. L., Person, G. et al. 2011. Marker-assisted selection to pyramid nematode resistance and the high oleic trait in peanut. *Plant Genome* 4: 110–117. https://doi.org/10.3835/plantgenome2011.01.0001.

Clevenger, J., Chu, Y., Chavarro, C., Agarwal, G., Bertioli, D. J. et al. 2017. Genome-wide SNP genotyping resolves signatures of selection and tetrasomic recombination in peanut. *Mol. Plant* 10: 309–322.

Collins, V. P., Cantor, A. H., Pescatore, A. J., Straw, M. L., Ford, M. J. et al. 1997. Pearl millet in layer diets enhances egg yolk "n-32" fatty acids. *Poultry Sci.* 76: 326–330.

Crossa, J., de Los Campos, G., Pérez, P., Gianola, D., Burgueño, J. et al. 2010. Prediction of genetic values of quantitative traits in plant breeding using pedigree and molecular markers. *Genetics* 186: 713–724.

Crossa, J., Pérez-Rodríguez, P., Cuevas, J., Montesinos-López, O., Jarquín, D. et al. 2017. Genomic selection in plant breeding: methods, models, and perspectives. *Trends Plant Sci.* 22: 961–975. https://doi.org/10.1016/j.tplants.2017.08.011.

Daetwyler, H. D., Villanueva, B. and Woolliams, J. A. 2008. Accuracy of predicting the genetic risk of disease using a genome-wide approach. *PLoS ONE* 3: e3395. https://doi.org/10.1371/journal.pone.0003395.

de los Campos, G., Naya, H., Gianola, D., Crossa, J., Legarra, A. et al. 2009. Predicting quantitative traits with regression models for dense molecular markers and pedigree. *Genetics* 182: 375–385. https://doi.org/10.1534/genetics.109.101501.

de los Campos, G., Gianola, D., Rosa, G. J., Weigel, K. A., Crossa, J. et al. 2010. Semi-parametric genomic-enabled prediction of genetic values using reproducing kernel Hilbert spaces methods. *Genetics Res.* 92: 295–308. https://doi.org/10. 1017/S0016672310000285.

de los Campos, G., Hickey, J. M., Pong-Wong, R., Daetwyler, H. D., Calus, M. P. et al. 2013. Whole-genome regression and prediction methods applied to plant and animal breeding. *Genetics* 193: 327–345.

Deokar, A. A., Ramsay, L., Sharpe, A. G., Diapari, M., Sindhu, A. et al. 2014. Genome wide SNP identification in chickpea for use in development of a high-density genetic map and improvement of chickpea reference genome assembly. *BMC Genomics* 15(1): 708.

Desta, Z. A. and Ortiz, R. 2014. Genomic selection: genome-wide prediction in plant improvement. *Trends Plant Sci.* 19: 592–601. https://doi.org/10.1016/j.tplants.2014.05.006.

Drucker, H., Burges, C. J. C., Kaufman, L., Smola, A., Vapnik, V. et al. 1997. Support vector regression machines. pp. 155–161. *In*: Mozer, M. C., Jordan, M. I. and Petsche, T. [eds.]. *Advances in Neural In-formation Processing System* 9, MIT Press, Cambridge, MA.

Dwivedi, S. L., Pande, S., Rao, J. N. and Nigam, S. N. 2002. Components of resistance to late leaf spot and rust among interspecific derivatives and their significance in a foliar disease resistance breeding in groundnut (*Arachis hypogaea* L.). *Euphytica* 125: 81–88.

Ejeta, G., Hansen, M. M. and Mertz, E. T. 1987. *In vitro* digestibility and amino acid composition of pearl millet (*Pennisetum typhoides*) and other cereals. *PNAS* (USA) 84: 6016–6019.

FAOSTAT. 2017. FAO Statistical Database, http://faostat.fao.org/ Accessed on 24th March 2020.

Furbank, R. T. and Tester, M. 2011. Phenomics–technologies to relieve the phenotyping bottleneck. *Trends Plant Sci.* 16: 635–644. https://doi.org/10.1016/j.tplants.2011.09.005.

Gianola, D. and van Kaam, J. B. C. H. M. 2008. Reproducing kernel Hilbert spaces regression methods for genomic assisted prediction of quantitative traits. *Genetics* 178: 2289–2303.

Govindaraj, M., Rai, K. N., Shanmugasundaram, P., Dwivedi, S. L., Sahrawat, K. L. et al. 2013. Combining ability and heterosis for grain iron and zinc densities in pearl millet. *Crop Sci.* 53: 507–517.

Grabiele, M., Chalup, L., Robledo, G. and Seijo, G. 2012. Genetic and geographic origin of domesticated peanut as evidenced by 5S rDNA and chloroplast DNA sequences. *Plant Syst. Evol.* 298: 1151–1165. https://doi.org/10.1007/s00606-012-0627-3.

Gupta, S., Nawaz, K., Parween, S., Roy, R., Sahu, K. et al. 2016. Draft genome sequence of *Cicer reticulatum* L., the wild progenitor of chickpea provides a resource for agronomic trait improvement. *DNA Res.* 24: 1–10. https://doi.org/10.1093/dnares/dsw042.

Gupta, S. K., Rai, K. N., Singh, P., Ameta, V. L., Gupta, S. K. et al. 2015. Seed set variability under high temperatures during flowering period in pearl millet (*Pennisetum glaucum* L. (R.) Br.). *Field Crop Res.* 171: 41–53.

Habier, D., Fernando, R. L., Kizilkaya, K. and Garrick, D. J. 2011. Extension of the Bayesian alphabet for genomic selection. *BMC Bioinformatics* 12: 186.

Hash, C. T., Yadav, R. S., Cavan, G. P., Howarth, C. J., Liu, H. et al. 2000. Marker-assisted backcrossing to improve terminal drought tolerance in pearl millet. pp. 114–119. *In*: Ribaut, J. M. and Poland, D. [eds.]. *Molecular Approaches for the Genetic Improvement of Cereals for Stable Production in Water Limited Environments: CYMMYT*, Batan MX.

Hash, C. T. and Witcombe, J. R. 2001. Pearl millet molecular marker research. *Int. Sorghum & Millets Newsletter*. 42: 8–15.

Hazel, L. and Lush, J. L. 1942. The efficiency of three methods of selection. *J. Hered.* 33: 393–399.

Heffner, E. L., Lorenz, A. J., Jannink, J. L. and Sorrells, M. E. 2010. Plant breeding with genomic selection: gain per unit time and cost. *Crop Sci.* 50: 1681–1690.

Heffner, E. L., Jannink, J. L., Iwata, H., Souza, E., Sorrells, M. E. et al. 2011. Genomic selection accuracy for grain quality traits in biparental wheat populations. *Crop Sci.* 51: 2597–2606.

Henderson, C. R. 1975. Best linear unbiased estimation and prediction under a selection model. *Biometrics* 31: 423–447. https://doi.org/10.2307/2529430.

Heslot, N., Yang, H. P., Sorrells, M. E. and Jannink, J. L. 2012. Genomic selection in plant breeding: a comparison of models. *Crop Sci.* 52: 146–160. https://doi.org/10.2135/cropsci2011.06.0297.

Hill, G. M. and Hanna, W. W. 1990. Nutritive characteristics of pearl millet grain in beef cattle diets. *J. Animal Sci.* 68: 2061–2066.

Hiremath, P. J., Kumar, A., Penmetsa, R. V., Farmer, A., Schlueter, J. A. et al. 2012. Large-scale development of cost-effective SNP marker assays for diversity assessment and genetic mapping in chickpea and comparative mapping in legumes. *Plant Biotechnol. J.* 10: 716–732. https://doi.org/10.1111/j.1467-7652.2012.00710.x.

Holland, J. B. 2004. Implementation of molecular markers for quantitative traits in breeding programs: Challenges and opportunities, 26. *In*: Fischer, T. et al. [eds.]. *New Directions for a Diverse Planet: Proc. for the 4th Int. Crop Science Congress, Brisbane, Australia.* 26 Sept.–1 Oct. 2004. Regional Institute, Gosford, Australia

Howard, R., Carriquiry, A. L. and Beavis, W. D. 2014. Parametric and nonparametric statistical methods for genomic selection of traits with additive and epistatic genetic architectures. *G3* (Bethesda) 4: 1027–1046. https://doi.org/10.1038/sj.tpj.6500167.

Jaganathan, D., Thudi, M., Kale, S., Azam, S., Roorkiwal, M. et al. 2015. Genotyping-by-sequencing based intra-specific genetic map refines a "*QTL hotspot*" region for drought tolerance in chickpea. *Mol. Genet. Genomics* 290: 559–571. https://doi.org/10.1007/s00438-014-0932-3.

Jain, M., Misra, G., Patel, R. K., Priya, P., Jhanwar, S. et al. 2013. A draft genome sequence of the pulse crop chickpea (*Cicer arietinum* L.). *Plant J.* 74: 715–729. https://doi.org/10.1111/tpj.12173.

Janila, P., Ramaiah, V., Rathore, A., Rupakula, A., Reddy, R. K. et al. 2013a. Genetic analysis of resistance to late leaf spot in interspecific groundnut. *Euphytica* 193: 13–25. https://doi.org/10.1007/s10681-013-0881-7.

Janila, P., Nigam, S. N., Pandey, M. K., Nagesh, P., Varshney, R. K. et al. 2013b. Groundnut improvement: use of genetic and genomic tools. *Front. Plant Sci.* 4: 23.

Janila, P., Pandey, M. K., Shasidhar, Y., Variath, M. T., Sriswathi, M. et al. 2016a. Molecular breeding for introgression of fatty acid desaturase mutant alleles (*ahFAD2A* and *ahFAD2B*) enhances oil quality in high and low oil containing peanut genotypes. *Plant Sci.* 242: 203–213. https://doi.org/10.1016/j.plantsci.2015.08.013.

Janila, P., Pandey, M. K., Manohar, S. S., Variath, M. T., Nallathambi, P. et al. 2016b. Foliar fungal disease-resistant introgression lines of groundnut (*Arachis hypogaea* L.) record higher pod and haulm yield in multilocation testing. *Plant Breeding* 135: 355–366. https://doi.org/10.1111/pbr.12358.

Jannink, J. L., Bink, M. C. and Jansen, R. C. 2001. Using complex plant pedigrees to map valuable genes. *Trends Plant Sci.* 6(8): 337–342.

Jannink, J. L., Lorenz, A. J. and Iwata, H. 2010. Genomic selection in plant breeding: from theory to practice. *Brief Funct. Genomics* 9: 166–177.

Jarquín, D., Crossa, J., Lacaze, X., Du Cheyron, P., Daucourt, J. et al. 2014. A reaction norm model for genomic selection using high-dimensional genomic and environmental data. *Theor. Appl. Genet.* 127: 595–607.

Jarquin, D., Howard, R., Liang, Z., Gupta, S. K., Schnable, J. C. et al. 2020. Enhancing hybrid prediction in pearl millet using genomic and/or multi-environment phenotypic information of inbreds. *Front. Genetics* 10: 1294.

Jonas, E. and de Koning, D. J. 2013. Does genomic selection has future in plant breeding? *Trends Biotechnol.* 31(9): 497–504.

Kanatti, A., Rai, K. N., Radhika, K., Govindaraj, M., Sahrawat, K. L. et al. 2014. Grain iron and zinc density in pearl millet: combining ability, heterosis and association with grain yield and grain size. *Springer Plus* 3: 763. https://doi.org/10.1186/2193-1801-3-763.

Khairwal, I. S., Rai, K. N., Andrews, D. J. and Harinarayana, G. 1999. *Pearl Millet Breeding.* Oxford and IBH Publishing, New Delhi

Khedikar, Y. P., Gowda, M. V. C., Sarvamangala, C., Patgar, K. V., Upadhyaya, H. D. et al. 2010. A QTL study on late leaf spot and rust revealed one major QTL for molecular breeding for rust resistance in groundnut (*Arachis hypogaea* L.). *Theor. Appl. Genet.* 121: 971–984. https://doi.org/10.1007/s00122-010-1366-x.

Kochert, G., Stalker, H. T., Gimenes, M., Galgaro, L., Lopes, C. R. et al. 1996. RFLP and cytogenetic evidence on the origin and evolution of allotetraploid domesticated peanut, *Arachis hypogaea* (leguminosae). *Am. J. Bot.* 83: 1282–1291. https://doi.org/10.1002/j.1537-2197.1996.tb13912.x.

Kolekar, R. M., Sujay, V., Kenta, S., Sukruth, M., Khedikar, Y. P. et al. 2016. QTL mapping for late leaf spot and rust resistance using an improved genetic map and extensive phenotypic data on a recombinant inbred line population in peanut (*Arachis hypogaea* L.). *Euphytica* 209: 147–156. https://doi.org/10.1007/s10681-016-1651-0.

Krishnamurthy, L., Vadez, V., Jyotsna Devi, M., Serraj, R., Nigam, S. N. et al. 2007. Variation in transpiration efficiency and its related traits in groundnut (*Arachis hypogaea* L.) mapping population. *Field Crops Res.* 103: 189–197.

Kulkarni, V. N., Rai, K. N., Dakheel, A. J., Ibrahim, M., Hebbara, M. et al. 2006. Pearl millet germplasm adapted to saline conditions. *Int. Sorghum Millets Newsletter* 47: 103–106.

Lande, R. and Thompson, R. 1990. Efficiency of marker-assisted selection in the improvement of quantitative traits. *Genetics* 124: 743–756.

Leal-Bertioli, S. C., José, A. C. V., Alves-Freitas, D. M., Moretzsohn, M. C., Guimarães, P. M. et al. 2009. Identification of candidate genome regions controlling disease resistance in *Arachis*. *BMC Plant Biol.* 9: 112.

Lee, M. 1995. DNA markers and plant breeding programs. *Adv. Agron.* 55: 265–344.

Li, Y., Ruperao, P., Batley, J., Edwards, D., Khan, T. et al. 2018. Investigating drought tolerance in chickpea using genome-wide association mapping and genomic selection on whole-genome resequencing data. *Front. Plant Sci.* 9: 190. https://doi.org/10.3389/fpls.2018.00190.

Liang, Z., Gupta, S. K., Yeh, C. T., Zhang, Y., Ngu, D. W. et al. 2018. Phenotypic data from inbred parents can improve genomic prediction in pearl millet hybrids. *G3: Genes, Genomes, Genetics* 8: 2513–2522.

Liao, B. S. 2014. Peanut breeding. pp. 61–78. *In*: Mallikarjuna, N. and Varshney, R. K. [eds.]. *Genetics, Genomics and Breeding of Peanut.* CRC Press, Boca Raton, USA.

Longin, C. F., Mi, X. and Würschum, T. 2015. Genomic selection in wheat: optimum allocation of test resources and comparison of breeding strategies for line and hybrid breeding. *Theor. Appl. Genet.* 128: 1297–1306.

Massman, J. M., Jung, H. J. G. and Bernardo, R. 2013. Genome wide selection versus marker-assisted recurrent selection to improve grain yield and stover-quality traits for cellulosic ethanol in maize. *Crop Sci.* 53: 58–66.

Mayor, P. J. and Bernardo, R. 2009. Genome wide selection and marker-assisted recurrent selection in doubled haploid versus F_2 populations. *Crop Sci.* 49(5): 1719–1725.

Meuwissen, T. H. E., Hayes, B. J. and Goddard, M. E. 2001. Prediction of total genetic value using genome wide dense marker maps. *Genetics* 157: 1819–1829.

Miller, M. R., Dunham, J. P., Amores, A., Cresko, W. A. and Johnson, E. A. et al. 2007. Rapid and cost-effective polymorphism identification and genotyping using restriction site associated DNA (RAD) markers. *Genome Res.* 17(2): 240–248. doi:10.1101/gr.5681207.
Moretzsohn, M. C., Gouvea, E. G., Inglis, P. W., Leal-Bertioli, S. C., Valls, J. F. et al. 2013. A study of the relationships of cultivated peanut (*Arachis hypogaea*) and its most closely related wild species using intron sequences and microsatellite markers. *Ann. Bot.* 111: 113–126. https://doi.org/10.1093/aob/mcs237.
Nakaya, A. and Isobe, S. N. 2012. Will genomic selection be a practical method for plant breeding? *Ann. Bot.* 110: 1303–1316. https://doi.org/10.1093/aob/mcs109.
Nigam, S. N., Chaudhari, S., Deevi, K. C., Saxena, K. B. and Janila, P. 2021. Trends in legume production and future outlook. pp. 7–48. In Genetic Enhancement in Major Food Legumes.
Oakey, H., Cullis, B., Thompson, R., Comadran, J., Halpin, C. et al. 2016. Genomic selection in multi-environment crop trials. *G3: Genes, Genomes, Genetics* 6(5): 1313–1326.
Ott, A., Liu, S., Schnable, J. C., Yeh, C. T. E., Wang, K. S. et al. 2017. tGBS® genotyping-by-sequencing enables reliable genotyping of heterozygous loci. *Nucleic Acids Res.* 45(21): e178–e178. doi: 10.1093/ nar/gkx853.
Pandey, M. K., Monyo, E., Ozias-Akins, P., Liang, X., Guimarães, P. et al. 2012. Advances in Arachis genomics for peanut improvement. *Biotechnol. Adv.* 30: 639–651.
Pandey, M. K., Roorkiwal, M., Singh, V., Lingam, A., Kudapa, H. et al. 2016. Emerging genomic tools for legume breeding: current status and future perspectives. *Front. Plant Sci.* 7: 455.
Pandey, M. K., Agarwal, G., Kale, S. M., Clevenger, J., Nayak, S. N. et al. 2017. Development and evaluation of a high density genotyping '*Axiom_Arachis*' array with 58K SNPs for accelerating genetics and breeding in groundnut. *Scientific Reports* 7: 40577.
Pandey, M. K., Chaudhari, S., Jarquin, D., Janila, P., Crossa, J. et al. 2020a. Genome-based trait prediction in multi-environment breeding trials in groundnut. *Theor. Appl. Genet.* 33(11): 3101–3117.
Pandey, M. K., Pandey, A. K., Kumar, R., Nwosu, V., Guo, B. et al. 2020b. Translational genomics for achieving higher genetic gains in post-genome era in groundnut. *Theor. Appl. Genet.* 133: 1679–1702.
Park, T. and Casella, G. 2008. The bayesian lasso. *J. Am. Stat. Assoc.* 103(482): 681–686.
Pérez-Rodríguez, P., Crossa, J., Bondalapati, K., Meyer, G. D. et al. 2015. A pedigree-based reaction norm model for prediction of cotton yield in multi-environment trials. *Crop Sci.* 55: 1143–1151.
Poland, J. and Rife, T. W. 2012. Genotyping-by-sequencing for plant breeding and genetics. *Plant Genome* 5: 92–102. https://doi.org/10.3835/plantgenome2012.05.0005.
Prasad, P. V., Kakani, V. G. and Upadhyaya, H. D. 2010. Growth and production of groundnut. *UNESCO Encyclopedia* pp. 1–26.
Rafalski, A. 2002. Applications of single nucleotide polymorphisms in crop genetics. *Curr. Opin. Plant Biol.* 5: 94–100.
Rai, K. N., Murty, D. S., Andrews, D. J. and Bramel-Cox, P. J. 1999. Genetic enhancement of pearl millet and sorghum for the semi-arid tropics of Asia and Africa. *Genome* 42(4): 617–628.
Rai, K. N., Kumar, K. A., Andrews, D. J. and Rao, A. S. 2001. Commercial viability of alternative cytoplasmic-nuclear male-sterility systems in pearl millet. *Euphytica* 121(1): 107–114.
Rai, K. N., Govindaraj, M. and Rao, A. S. 2012. Genetic enhancement of grain iron and zinc content in pearl millet. *Qual. Assur. Saf. Crop. Foods* 4: 119–125.
Rai, K. N., Patil, H. T., Yadav, O. P., Govindaraj, M., Khairwal, I. S. et al. 2014. Variety Dhanashakti (Pearlmillet). *Indian J. Genet. Plant Breed.* 74(3): 405–406.

Riedelsheimer, C., Czedik-Eysenberg, A., Grieder, C., Lisec, J., Technow, F. et al. 2012. Genomic and metabolic prediction of complex heterotic traits in hybrid maize. *Nat. Genet.* 44: 217–220.

Roorkiwal, M., Sawargaonkar, S. L., Chitikineni, A., Thudi, M., Saxena, R. K. et al. 2013. Single nucleotide polymorphism genotyping for breeding and genetics applications in chickpea and pigeonpea using the BeadXpress platform. *Plant Genome* 6: 2. https://doi.org/10.3835/plantgenome2013.05.0017.

Roorkiwal, M., Rathore, A., Das, R. R., Singh, M. K., Jain, A. S. et al. 2016. Genome-enabled prediction models for yield related traits in chickpea. *Front. Plant Sci.* 7: 1666.

Roorkiwal, M., Jarquin, D., Singh, M. K., Gaur, P. M., Bharadwaj, C. et al. 2018. Genomic-enabled prediction models using multi-environment trials to estimate the effect of genotype × environment interaction on prediction accuracy in chickpea. *Scientific Reports* 8: 11701. https://doi.org/10.1038/s41598-018-30027-2.

Ruperao, P., Chan, C. K., Azam, S., Karafiátová, M. et al. 2014. A chromosomal genomics approach to assess and validate the desi and kabuli draft chickpea genome assemblies. *Plant Biotechnol. J.* 12: 778–786. https://doi.org/10.1111/pbi.12182.

Rutkoski, J., Benson, J., Jia, Y., Brown-Guedira, G., Jannink, J. L. et al. 2012. Evaluation of genomic prediction methods for Fusarium head blight resistance in wheat. *The Plant Genome* 5(2): 51–61.

Schön, C., Utz, S., Groh, B., Truberg, S., Openshaw, S. et al. 2004. QTL mapping based on resampling in a vast maize testcross experiment confirms the infinitesimal model of quantitative genetics for complex traits. *Genetics* 167: 485–498.

Sehgal, D., Skot, L., Singh, R., Srivastava, R. K., Das, S. P. et al. 2015. Exploring potential of pearl millet germplasm association panel for association mapping of drought tolerance traits. *PLoS ONE* 10(5): e0122165. https://doi.org/10.1371/journal.pone.0122165.

Shikha, M., Kanika, A., Rao, A. R., Mallikarjuna, M. G., Gupta, H. S. et al. 2017. Genomic selection for drought tolerance using genome-wide SNPs in maize. *Front. Plant Sci.* 8: 550.

Singh, A. K., Subrahmanyam, P. and Moss, J. P. 1984. The dominant nature of resistance to *Puccinia ara chidis* in certain wild *Arachis* species. *Oleagineux* 39: 535–537.

Spindel, J., Begum, H., Akdemir, D., Virk, P., Collard, B. et al. 2015. Genomic selection and association mapping in rice (*Oryza sativa*): effect of trait genetic architecture, training population composition, marker number and statistical model on accuracy of rice genomic selection in elite, tropical rice breeding lines. *PLoS Genet.* 11(2): e10046982. https://doi.org/10.1371/journal.pgen.1004982.

Srivastava, R. K., Singh, R. B., Pujarula, V. L., Bollam, S., Pusuluri, M. et al. 2020. Genome-wide association studies and genomic selection in Pearl Millet: Advances and prospects. *Front. Genet.* 10: 1389.

Sujay, V., Gowda, M. V. C., Pandey, M. K., Bhat, R. S., Khedikar, Y. P. et al. 2012. Quantitative trait locus analysis and construction of consensus genetic map for foliar disease resistance based on two recombinant inbred line populations in cultivated groundnut (*Arachis hypogaea* L.). *Mol. Breed.* 30: 773–788. https://doi.org/10.1007/s11032-011-9661-z.

Technow, F., Burger, A. and Melchinger, A. E. 2013. Genomic prediction of northern corn leaf blight resistance in maize with combined or separated training sets for heterotic groups. *G3: Genes, Genomes, Genetics* 3: 197–203. https://doi.org/10.1534/g3.112.004630.

Tester, M. and Langridge, P. 2010. Breeding technologies to increase crop production in a changing world. *Science* 327: 818–822. https://doi.org/10.1126/science.1183700.

Thudi, M., Bohra, A., Nayak, S. N., Varghese, N., Shah, T. M. et al. 2011. Novel SSR markers from BAC-end sequences, DArT arrays and a comprehensive genetic map with 1,291 marker loci for chickpea (*Cicer arietinum* L.). *PLoS ONE* 6: e27275. https://doi.org/10.1371/journal.pone.0027275.

Tibshirani, R. 1996. Regression shrinkage and selection via the LASSO. *J. R. Stat. Soc.: Series B.* 58: 267–288.
Varshney, R. K., Mohan, S. M., Gaur, P. M., Gangarao, N. V. P. R., Pandey, M. K. et al. 2013a. Achievements and prospects of genomics-assisted breeding in three legume crops of the semi-arid tropics. *Biotechnol. Adv.* 31: 1120–1134.
Varshney, R. K., Song, C., Saxena, R. K., Azam, S., Yu, S. et al. 2013b. Draft genome sequence of chickpea (*Cicer arietinum*) provides a resource for trait improvement. *Nat. Biotechnol.* 31(3): 240–246. https://doi.org/10.1038/nbt.2491.
Varshney, R. K., Gaur, P. M., Chamarthi, S. K., Krishnamurthy, L., Tripathi, S. et al. 2013c. Fast-Track Introgression of "*QTL-hotspot*" for root traits and other drought tolerance traits in JG 11, an elite and leading variety of chickpea. *The Plant Genome* 6: 07. https://doi.org/10.3835/plantgenome2013.07.0022.
Varshney, R. K., Thudi, M., Nayak, S. N., Gaur, P. M., Kashiwagi, J. et al. 2014. Genetic dissection of drought tolerance in chickpea (*Cicer arietinum* L.). *Theor. Appl. Genet.* 127(2): 445–462.
Varshney, R. K. 2016. Exciting journey of 10 years from genomes to fields and markets: some success stories of genomics-assisted breeding in chickpea pigeonpea and groundnut. *Plant Sci.* 242: 98–107.
Varshney, R. K., Shi, C., Thudi, M., Mariac, C., Wallace, J. et al. 2017. Pearl millet genome sequence provides a resource to improve agronomic traits in arid environments. *Nat. Biotechnol.* 35(10): 969–976.
Varshney, R. K., Pandey, M. K., Bohra, A., Singh, V. K., Thudi, M. et al. 2019. Toward sequence-based breeding in legumes in the post-genome sequencing era. *Theor. Appl. Genet.* 132(3): 797–816.
Velu, G., Rai, K. N., Sahrawat, K. L. and Sumalini, K. 2008a. Variability for grain iron and zinc contents in pearl millet hybrids. *J. SAT Agric. Res.* 6: 1–4.
Velu, G., Rai, K. N. and Sahrawat, K. L. 2008b. Variability for grain iron and zinc content in a diverse range of pearl millet populations. *Crop Improv.* 35(2): 186–191.
Verma, S., Gupta, S., Bandhiwal, N., Kumar, T., Bharadwaj, C. et al. 2015. High-density linkage map construction and mapping of seed trait QTLs in chickpea (*Cicer arietinum* L.) using Genotyping-by-Sequencing (GBS). *Scientific Reports* 5: 17512.
Vindhiyavarman, P., Raveendran, T. S. and Ganapathi, T. 1993. Inheritance of rust resistance in groundnut. *Madras Agri. J.* 80: 175–176.
Wang, N., Liu, B., Liang, X., Zhou, Y., Song, J. et al. 2019. Genome-wide association study and genomic prediction analyses of drought stress tolerance in China in a collection of off-PVP maize inbred lines. *Mol. Breed.* 39: 113.
Yadav, O. P., Rai, K. N., Rajpurohit, B. S., Hash, C. T., Mahala, R. S. et al. 2012a. Twenty-five Years of Pearl Millet Improvement in India. All India Coordinated Pearl Millet Improvement Project Jodhpur, India.
Yadav, O. P., Rai, K. N. and Gupta, S. K. 2012b. Pearl millet: genetic improvement for tolerance to abiotic stresses. pp. 261–288. *In*: Tuteja, N., Gill, S. S. and Tuteja, R. [eds.]. *Improving Crop Productivity in Sustainable Agriculture.* First Edition. Wiley-VCH Verlag GmbH & Co. KGaA, Weinheim.
Yadav, H. P., Gupta, S. K., Rajpurohit, B. S. and Pareek, N. 2016. Pearl millet pp. 205–224. *In*: Singh, M. and Kumar, S. [eds.]. *Broadening the Genetic Base of Grain Cereals.* First Edition. Springer India Private Ltd, New Delhi, India.
Zegada-Lizarazu, W. and Iijima, M. 2005. Deep root water uptake ability and water use efficiency of pearl millet in comparison to other millet species. *Plant Prod. Sci.* 8: 454–460.
Zhang, J., Song, Q., Cregan, P. B. and Jiang, G. L. 2016. Genome-wide association study, genomic prediction and marker-assisted selection for seed weight in soybean (*Glycine max*). *Theor. Appl. Genet.* 129: 117–130. https://doi.org/10.1007/s00122-015-2614-x.

Zhao, Y., Zhang, C., Chen, H., Yuan, M., Nipper, R. et al. 2016. QTL mapping for bacterial wilt resistance in peanut (*Arachis hypogaea* L.). *Mol. Breed.* 36: 13. https://doi.org/10.1007/s11032-015-0432-0.

Zhong, S., Dekkers, J. C., Fernando, R. L. and Jannink, J. L. 2009. Factors affecting accuracy from genomic selection in populations derived from multiple inbred lines: a barley case study. *Genetics* 182(1): 355–364.

Zhuang, W., Chen, H., Yang, M., Wang, J., Pandey, M. K., Zhang, C. et al. 2019. The genome of cultivated peanut provides insight into legume karyotypes, polyploid evolution and crop domestication. *Nat. Genet*. 51: 865–876.

6

Genomic Selection in *Solanaceae*: Status, Opportunities and Future Prospects

Vandana Jaiswal,[1,2,*] *Vijay Gahlaut,*[1] *Sushil Satish Chhapekar*[3] and *Nirala Ramchiary*[3,*]

ABSTRACT

Solanaceae family contains major vegetable crop plants such as *Capsicum*, tomato, potato, and eggplant. Although conventional breeding has achieved considerable success in breeding these crops, however, the advancement of molecular genetics and genomics techniques have greatly supplemented the development of high yielding, rich in nutrients, tolerant to biotic and abiotic stress crop varieties in shorter time periods. Genomic selection (GS) is one of the high throughput emerging techniques of plant breeding that has been successfully applied for the selection of several complex traits in many crop plants. In *Solanaceae*, GS has been conducted for traits such as fruit size, disease resistance and other traits mostly in tomato. However, in case of other *Solanaceae* crops like *Capsicum*, eggplant, and potato there is very limited information on GS and therefore needs attention. In this chapter, an attempt has been made to compile the genomic resources

[1] Division of Biotechnology, CSIR-Institute of Himalayan Bioresource and Technology, Palampur 176061, India.

[2] Academy of Scientific and Innovative Research (AcSIR), CSIR-IHBT, Palampur, Himachal Pradesh, 176061, India.

[3] School of Life Sciences, Jawaharlal Nehru University, New Delhi 110067.

* Corresponding authors

available on *Solanaceae* crop plants which can be utilized for GS and listed limited available study reports on GS on those crop plants. Further, we explored methodologies and strategies of GS that are being applied in different crops which can also be adopted towards improvement of *Solanaceae.* Furthermore, advantages and future scope of adopting GS for the development of better varieties in *Solanaceae* crops has been discussed.

Introduction

Solanaceae is one of the largest family of flowering crops which contains 97 genera and approximately 2700 species (Olmsteads et al., 2008; Sarkinen et al., 2013). Plants from this family are also called nightshade plants. Tomato, potato, eggplant and chili peppers, which are the most important horticultural crops cultivated and consumed worldwide, belong to this family. Solanaceous crops have a diverse origin. For example, tomato and pepper originated from Central and South America while eggplant originated from Indo-China. These crops have high nutritional and economic value. If we talk about food consumption, potato is the 5th largest crop globally and tomato, eggplant and pepper ranked 9th, 26th, 33rd respectively (FAO, 2013). Among the vegetable crops, potato is the largest, and tomato is the second largest crop consumed worldwide. *Solanaceae* crops are present in the top ten produces in the world, where tomato ranked 4th with $58B crop value and potato ranked 8th with $49B crop value (FAO 2013). The total production of tomato in 2017 was 182 million tons which equals to more than $60 billion.

Tomato is one of the most important vegetable crops of the family *Solanaceae.* This crop is easy to cultivate, annual, early flowering, and autogamous. It is a good source of vitamins (A and C) and has very high lycopene content with antioxidant properties (Fentik et al., 2017) and other compounds beneficial to health. Tomato is also being used by researchers as a model crop plant for studying various trait phenotypes and their underlying genes using genetics, genomics, and molecular biology tools and techniques. Like tomato, pepper is also an important cash crop which is rich in vitamins, minerals, antioxidants and metabolites. Pepper is mostly used as a spice, although few non pungent varieties are used as vegetables, and consumed raw or cooked. Pepper also has several medicinal values like antimicrobial, anti-inflammatory and anti-cancerous affects (Jaiswal et al., 2019). Similarly, eggplant is very important for human health particularly for heart related issues. Eggplant is a potential reservoir of fibres, vitamin C, vitamin B-6, minerals such as copper, manganese, potassium, and antioxidants which are necessary elements to keep our hearts healthy (Gurbuz et al., 2018).

Considering the potential value in combating malnutrition, there have been continuous breeding efforts for the improvement of *Solanaceae* crops.

The advancements in modern genetic and genomic tools helped to develop several marker systems in many solanaceous crops which significantly complemented rapid trait breeding programs of these crops (Brachi et al., 2011; Sim et al., 2012; Cheng et al., 2016). Using molecular markers, a large number of genes/QTLs associated with several agriculturally important traits have been identified in *Solanaceae* crops. Marker-assisted selection is also being performed for the betterment of varieties. As most of the economically important traits are quantitative traits and controlled by a number of minor genes, breeding for such complex traits is very difficult. However, the use of recently developed high throughput genomics technologies have been helpful in genomic selection (GS) of important traits and proved to be very useful which have shown a significant increase of the genetic gain per cycle as compared to the conventional breeding and marker assisted selection (MAS). GS also reduces the cost as well as labour for phenotyping and has been applied in several crops like wheat, rice, chickpeas, pearl millet and others (Roorkiwal et al., 2016; Xu et al., 2018; Liang et al., 2018; Sweeney et al., 2019). Although among the *Solanaceae* crops, study reports on GS have been found mostly in tomato. In this chapter, an attempt has been made to compile available information on genetic advances, and the potential application of GS and associated models, tools and techniques in four major *Solanaceae* crops tomato, pepper, eggplant and potato.

Genetic resources in *Solanaceae*

Plant genetic resources are important for maintaining diversity of crop plants to be used in breeding programs and for their conservation. For *Solanaceae* crops, Asian Vegetable Research and Development Center (AVRDC) gene bank in Taiwan, has a large number of germplasms which includes 8170 accessions of pepper, 8151 of tomato and 3702 accessions of eggplant, as well as thousands of germplasms from indigenous vegetables (https://avrdc.org/about-avrdc/history/2010s/). The United States Department of Agriculture (USDA) through the Germplasm Resources Information Network (GRIN) has collected and maintained a total of 775 accessions of eggplant, 8418 of tomato, 1025 of potato and 5029 of peppers. These organisations are principally involved in collection, multiplication, and distribution of seeds to researchers all over the world. Apart from this, several public and private organizations in different countries have been involved in germplasm collection, multiplication and conservation. One of them is the European Cooperative Programme for Plant Genetic Resources (ECPGR), which conserves the genetic resources in long term storage conditions for sustainable utilization.

Whole genome sequencing

Among the *Solanaceae* crops, the first whole genome sequencing was done for tomato (The Tomato Genome Consortium, TCG, 2012) using cultivar 'Heinz 1706'. A combination of different next generation sequencing (NGS) technologies (Illumina Genome Analyser, SOLID sequencing and 454/Roche GS FLX) was utilized for whole genome sequencing. A genome size of 900 MB was predicted which included 91 scaffolds mapped on 12 chromosomes. Further, a total of 96 conserved miRNA genes were also reported. The genome analyses suggested the increase of members in gene families by triplication events that are involved in regulation of fruit traits such as texture, colour, taste, fleshiness among others. Further, re-sequencing of 84 accessions of tomato for the purpose of studying the genetic variability existing between cultivated tomato and its wild progenitor was also reported (Aflitos et al., 2014).

The draft genome sequence of *Capsicum annum* cv. CM334 was reported by Kim et al. (2014). The variety CM334 was chosen for sequencing because this cultivar was reported to show resistance to multiple diseases. A total of 650.2 Gb data with 186.6 × genome coverage was generated using the Illumina NGS platform. Genome size was estimated to be 3.48 Gb. Anchoring based on high density map could assemble 86% (2.63 Gb; 1357 scaffolds) of the sequences on 12 chromosomes of *Capsicum* genome. In the 2.63 Gb assembled genome, a total of 34,903 protein coding genes were identified which were highly similar to tomato genes. Further, it has been observed that genome size is approximately four times larger than tomato which could be attributed to the presence of a considerable number of repetitive elements like LTR retrotransposons (Kim et al., 2014) in the *Capsicum* genome. In the same year, to investigate the evolution and domestication pattern, whole genome sequencing was done for cultivated Zunla-1 (*C. annuum* L.) and wild Chiltepin (*C. annuum* var. glabriusculum) (Qin et al., 2014) and reported genome sizes of 3.26 Gb and 3.07 Gb, respectively. They reported a total of 6,527 long noncoding (lnc) RNAs (5,976 intergenic, 222 intron-overlapping) and 5,581 phased siRNAs were also identified. Furthermore, they identified a total of 176 miRNAs (including 35 pepper specific miRNAs). Candidate genes for various traits have also been predicted using a gene annotation tool.

Reference genome sequence/datasets for eggplant were made available in 2014 (Hirakawa et al., 2014). Eggplant variety 'Nakate-Shinkuro' was sequenced using Illumina platform (HiSeq 2000 sequencer) and led to the generation of 1127 Mb genome size including 1,321,157 scaffolds. Further, they also reported transcriptome analysis and identified a total 42,035 genes in eggplant, out of which 16,573 were orthologous to the tomato genes. Furthermore, the conserved syntenic blocks and orthologous genes of eggplants with tomato involved governing fruit traits and disease resistance.

Like tomato, potato genome was sequenced by the international community under the collaborative umbrella project named 'The Potato Genome Sequencing Consortium (PGSC) project' using Illumina Genome Analyser and Roche Pyrosequencing Technologies (Potato Genome Sequencing Consortium, 2011). To account for the high heterozygosity, two double monoploid potato clones were used for sequencing. Using the whole genome sequencing approach, 86% of the sequenced genome, i.e., 844 Mb potato genome was assembled which led to the prediction of 39,031 protein coding genes located on 12 chromosomes. Among these 2,642 and 3,372 genes were asteroid-specific and potato lineage-specific.

Important genes/QTLs involved in stress tolerance and economic traits

A number of gene/QTLs have been reported for stress tolerance and agronomic traits in *Solanaceae* crop (Table 1) that may play a vital role in its improvement. For instance, *patatin* gene found involved in the biosynthesis of major storage protein (Ganal et al., 1991). *AUX/IAA*, *SAUR* (Small Auxin Up-regulated RNAs), *CaM* (Calmodulin) genes plays a significant role in abiotic stresses like temperature, water and salt stress (Wu et al., 2012a, b; Zhao et al., 2013). For biotic stresses, *R loci* and *Glabodera* resistance loci (Grube et al., 2000) controlling the resistance against Potato Virus Y (PVY), Tomato Spotted Wilt Virus (TSWV), Tobacco Mosaic Virus (TMV) and *Glabodera* spp. were reported. For viral resistance, the common resistance loci were detected containing two potyvirus R genes (*Pvr4* and *Pvr7*) and *Tsw* gene (Lefebvre et al., 2002; Thabuis et al., 2004). For resistance to diseases such as *Phytophthora infestans* and late blight, QTLs have been identified that are colocalized among two or more species of *Solanaceae* (Collins et al., 1999; Ghislain et al., 2001; Brouwer et al., 2004). For nematode tolerance, genes such as *Mi3, eIF4E* and *Gpa2* have been identified (Dijan-Caporalino et al., 2001; Ruffel et al., 2006).

Whole genome comparative QTL analyses for traits such as fruit weight, pericarp thickness, and fruit shape in tomato and pepper have also been done (Van der knaap et al., 2003; Ben-Chaim et al., 2006). For example, in eggplant QTLs *fw2.1*, *fw9.1* and *fw11.1* were found to be orthologous to the tomato QTLs *fw2.2 fw9.2* and *fw11.1*, respectively (Grandillo et al., 1999; Frary et al., 2000; Doagnlar et al., 2002b). Similarly, three fruit shape related QTLs were found conserved among *Solanaceae* during domestication (Grandillo et al., 1999; Ben Chaim et al., 2003a, b). Cloning and characterization of *OVATE* like genes (associated with the fruit shape) also suggested the conservancy of the gene in *Solanaceae* (Tsaballa et al., 2012). Major fruit

Table 1. List of QTL mapping studies conducted in major vegetable crops of *Solanaceae* family.

Crop	Trait	Mapping population[@]	Marker[#]	Reference
Tomato	Fruit quality	RILs	RAPD, RFLP, AFLP	Causse et al., 2002
	Tomato volatiles	Natural accessions	SSRs	Zhang et al., 2015
	Leaf traits	BILs	SNPs	Fulop et al., 2016
	Fruit mineral contents	RILs	SSR, SNP, InDel	Capel et al., 2017
	Fruit quality	BILs	SNPs	Celik et al., 2017
	Resistance to late blight	F_2 and derived families	SNPs	Panthee et al., 2017
	Fruit nutrition and flavour	RILs	SSR, SNPs	Kimbara et al., 2018
	Shape of type VI glandular trichome	BC	SNP, CAPS	Bennewitz et al., 2018
	Heat Tolerance	F_2	SSR, InDel	Wen et al., 2019
	Fruit size and yield	F_2	SNP	Brekke et al., 2019
Pepper	Plant development and fruit traits	RILs	AFLPs, SSRs, RFLPs, SSAP, STS	Brachi et al., 2009
	Growth traits	DHs	SSRs, AFLPs	Mimura et al., 2010
	Horticultural traits	RILs	SNPs	Han et al., 2016
	First flower node	RILs	SLAF	Zhang et al., 2019
Eggplant	Fruit shape and colour development	F_2	RAPD, AFLP	Nunome et al., 2001
	Morphological traits	F_2	RFLPs	Frary et al., 2003
	Agronomic traits	F_2	SNPs, HRMs, CAPSs, RFLPs, SSRs and COSII	Portis et al., 2014
	Morphological and physiological traits	F_2	SNPs, HRMs, CAPSs, RFLPs, SSRs and COSII	Toppino et al., 2016

Potato	Economic traits	F_1	SNPs	Massa et al., 2015
	Glucose concentration, processing quality, vine maturity, and agronomic traits	F_1	SNPs	Massa et al., 2018
	Tuber shape, flesh and skin color	F_2	SNPs	Meijer et al., 2018
	Tuber morphology	Full sibling progenies	DArT, CAPS, SCAR	Hara-Skrzypiec et al., 2018
	Disease resistance	F_1	SNPs	Santa et al., 2018
	Early maturity traits	F_1	SNPs, SSRs	Li et al., 2018
	Tuber starch content and plant maturity	F_1	SSRs, AFLPs	Li et al., 2019

@ RILs = recombinant inbred lines, BILs = backcross inbred lines, BC = backcross population, DHs = double haploids.

RAPD = randomly amplified polymorphic DNA, RFLP = restriction fragment length polymorphism, AFLP = amplified fragment length polymorphism, SSR = simple sequence repeat, SNP = single nucleotide polymorphism, InDel = insertion deletion, SSAP = sequence specific amplified polymorphism, STS = sequence tag site, SLAF = sequence locus amplified fragment, DarT = diversity array technology, CAPS = cleave amplified polymorphic site, SCAR = sequence characterized amplified region.

shape genes such as *fw* are also found conserved in *Solanaceae* (Livingstone et al., 1999; Monforte et al., 2001; Yates et al., 2004; Zygier et al., 2005).

Candidate genes analysis governing pigmentation led to the identification of ten structural genes of involved carotenoid biosynthetic pathways in *Solanaceae* plants (Thorup et al., 2000). The important genes such as *Ccs (capsanthin capsorubin synthase), lycopene-β-cyclase* gene *Zds* loci responsible for ζ-Carotene desaturase, *Ccs* and the *B* locus responsible for hyper-accumulation of b-carotene were mapped in tomato as well in *Capsicum* (Thorup et al., 2000). Besides, carotenoid biosynthesis, genes involved in anthocyanin (another important pigment of *Solanaceae*) biosynthesis have also been identified (De Jong et al., 2004). Transcription factors playing an important role in pigment biosynthesis are also reported in *Solanaceae*. For example, *GLK2* (GOLDEN2-like transcription factor) is reported to be involved in fruit colour development and ripening in pepper and tomato (Powell et al., 2012; Nguyen et al., 2014; Brand et al., 2014; Nadakuduti et al., 2014). Again, in tomato and pepper Pan et al. (2013) identified an allied but different transcription factor named, APRR2-like (Arabidopsis Pseudo Response Regulator 2 Like) gene having a similar function with tomato *SlGLK2*. List of genes and QTLs identified in Solananceae plants involved in governing several economically important traits are listed in Table 1.

Marker-assisted selection (MAS) in *Solanaceae* family

The development of genomic tools (like markers, statistical tools) helps in identification of important genes and QTLs; which can be transferred through MAS to the existing well adapted varieties from the donor genotypes for further improvement. MAS has been conducted in a number of major crops such as wheat, rice, soybean, millets and extensively reviewed from time to time. Reports for MAS are also available in the *Solanaceae* family including tomato, *Capsicum*, potato and eggplant (Table 2). In case of *Solanaceae*, MAS is mainly conducted for disease resistance which is controlled by a few major genes. For example, in tomato MAS has been conducted for introgression of several genes providing resistance against powdery mildew, fusarium vascular wilt, leaf curl virus, late blight, and bacterial wilt. Similarly, in case of potato, MAS has been conducted to enhance the resistance against Columbia root-knot nematode, *Phytopthora infestans* and potato virus Y. Genes have been transferred to develop resistance against potato virus Y, tomato spotted wilt virus, pepper mild mottle virus and Verticillium wilt resistance in *Capsicum* and eggplant through MAS. Besides biotic stress, MAS has also been conducted for abiotic stresses in *Solanaceae*. For example, gene *SlCE1* responsible for cold stress tolerance has been transferred to existing lines to make cold stress tolerant tomato. Reports are also available on MAS for quality traits in

Table 2. Details of different MAS programs in *Solanaceae* family.

Crop	Disease/stress/trait	Genes/markers	References
Tomato	Powdery mildew	*ol-1*	Huang et al., 2000
	Fusarium vascular wilt	*I-2*	El Mohtar et al., 2007
	Cold stress	*SlICE1*	Miura et al., 2012
	Tomato leaf curl virus	*Ty-2, Ty-3*	Kumar et al., 2014 Prasanna et al., 2015
	Multiple disease resistant	*Ty-2, Ty-3* (Tomato yellow leaf curl disease); *Ph-2, Ph-3* (Late blight); *I-2* (Fusarium wilt); *Bwr-12* (Bacterial wilt)	Hanson et al., 2016
	Multiple disease resistant	*Ty-1, Ty-2 and Ty-3* (ToLCV), late blight (*Ph-2* and *Ph-3*) and root knot nematodes resistance (*Mi-1.2*)	Kumar et al., 2019
Potato	Columbia root-knot nematode	STS marker for $R_{Mc1(blb)}$	Zhang et al., 2007
	Phytophthora infestans resistance	$R_{Pi\text{-}mcd1}$ and $R_{Pi\text{-}ber}$	Tan et al., 2010
	Potato Virus Y Resistance	R_{yadg} and R_{ysto}	Fulladolsa et al., 2015
Pepper	Potato virus Y (PVY), Tomato spotted wilt virus (TSWV) and Pepper mild mottle virus (PMMoV)	*Pvr4, Tsw* and *L4*	Özkaynak et al., 2014
	Capsaicinoid content	*p-AMT* and *Pun1*	Tanaka et al., 2014
	Capsaicinoid content	*p-AMT*	Jeong et al., 2015
	Phytophthora capsici resistant	*RB* gene	Bagga et al., 2019
Eggplant	Verticillium wilt resistance	Allele-specific marker for the *Ve* gene	Liu et al., 2015

Capsicum where enhanced capsicinoid content in *C. annuum* spp. through the transfer of *AMT* gene was reported (Tanaka et al., 2014; Jeong et al., 2015).

Genomic selection

Crop breeders are facing an uphill task in keeping up with production due to ever increasing demands as a result of growing populations. The task is being further complicated by fast changing climates, limited water resources availability and low soil vigour. Considering these circumstances, conventional plant breeding needs to pick up pace. GS may play a vital role in current and future plant breeding to increase crop productivity and nutritional content (Varshney et al., 2017). GS reduces the breeding cycle through enhancement of the genetic gain per breeding cycle. Unlike MAS, GS explores the genome wide markers for estimation of breeding value which enables the capturing of even minor QTLs involved in complex phenotypes (Crossa et al., 2017; Varshney et al., 2017; Wang et al., 2018). With the advancement and cost-effective genotyping platforms, GS is found to be a very effective approach for crop improvement. Millions of molecular markers have been developed in each of the important crops including *Solanaceae* (Brachi et al., 2011; Sim et al., 2012; Cheng et al., 2016), and per reaction cost of genotyping is also very low and feasible for breeders which ultimately reduces the cost of genotyping. Further, GS also lowers the phenotyping efforts in every generation of breeding programs. In GS, phenotyping is being conducted in the training population for the estimation of the marker allele breeding value; and in breeding population selection is done based on the genome wide estimated breeding value (GEBV) through genotyping. Non requirement of phenotyping of the breeding population reduces the demand of land area and high cost of phenotyping. Further, breeding for roots or underground parts of the plant is very difficult with conventional plant breeding due to complex phenotyping efforts. In such a situation, GS is found to be a potential approach for crop improvement.

Fruits are the main products in major *Solanaceae* crops*'; hence breeding is* mainly done for fruit traits except for potato (Tanaka et al., 2014; Jeong et al., 2015). However, the nutritional quality of fruits also depends on the nutritional uptake by roots, thus breeding for root traits also affects the nutritional properties of fruits. Unfortunately, phenotyping for root traits as well as nutritional quality traits are quite tedious, time consuming and expensive; and in such a scenario GS would prove be a great help. The pipeline for GS which can be utilized in crop plants including *Solanaceae* is illustrated in Fig. 1. GS uses two types of populations which are called training and breeding populations respectively. The training population is composed of diverse germplasms which are genotyped with high density markers and phenotyped. Using genotyping and phenotyping breeding value is calculated and used to estimate genomic estimated breeding value (GEBV) in the breeding population where selections are made. GS is an advanced breeding technique which complements the conventional breeding, and integration of both can accelerate the breeding program substantially in all the crop plants including *Solanaceae*.

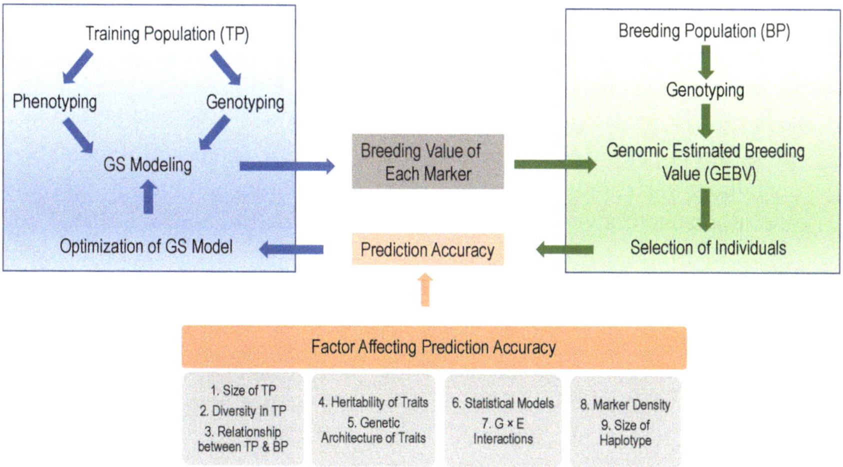

Fig. 1. Flow diagram illustrating blueprint of genomic selection in crop plants.

Factors affecting prediction accuracy in GS

In GS modelling, prediction accuracy is very important for the selection of genotypes. Prediction accuracy is affected by several factors. The following is a briefing on the influencing factors: –

i. *Size of the training population*; prediction accuracy is directly proportional to the size of the training population (Duangjit et al., 2016; Yammamoto et al., 2016; 2017). With the increase in the size of the training population, more genotypes will be genotyped and phenotyped; and with an increase in the sample size, error rate decreases, and prediction accuracy increases. Although, after reaching a certain population size, prediction accuracy becomes saturated and forms a plateau. Thus, we can say, an optimum training population size is required to maximize the prediction accuracy.

ii. *Diversity available in the training population*; genetic diversity of the training population also affects the prediction accuracy. More the training population diversity more is the prediction accuracy of GS models (Wang et al., 2018).

iii. *Relationship between training population and breeding population*; this factor also affects the prediction accuracy substantially; a close relationship results in higher accuracy (Saatchi et al., 2010; Akdemir et al., 2015; Isidro et al., 2015; Akdemir and Isidro, 2019).

iv. *Heritability of traits*; prediction accuracy also depends on the heritability of traits and generally found high in highly heritable traits (Wang et al., 2018); however, models are available to deal with limitations. For instance, GBLUP showed high accuracy for traits with low heritability which are controlled by several minor effect genes.

v. *Genetic architecture of trait*: genomic prediction accuracy is generally lower for highly complex phenotypes which are controlled by many small effect genes/QTLs. In such cases, with the increase in population size, more variations can be used to train the model and thereby increase prediction accuracy (Wang et al., 2018; Akdemir and Isidro, 2019).

vi. *Maker density used for estimation of GEBV*; high density markers often increase the prediction accuracy, however, a further increase is not achievable after reaching a prediction accuracy plateau (Crossa et al., 2017).

vii. *Size of haplotype*; when marker density is fixed, length of haplotype also affects the prediction accuracy, and should be optimum for the highest accuracy (Calus et al., 2008; Villumsen et al., 2009).

viii. *Statistical models*; in order to estimate GEBV, different assumptions are made in different statistical models which ultimately become an important factor for high prediction accuracy (de los Campos et al., 2013; Akdemir and Isidro, 2019).

Tools/software for genomic selection

Massive genotypic and phenotypic data are generated in GS, posing a great challenge for breeders to handle data. It requires appropriate computational infrastructure as well as expertise in statistics and bioinformatics. Thus, to overcome this limitation, and increase the acceptance of GS, different tools and software are being developed. The United States Department of Agriculture (USDA) database comprises information which facilitates GS in dairy cattle (Wiggans et al., 2010) wheat and barley under the Triticeae Coordinated Agricultural Project (T-CAP) (http://triticeaetoolbox.org/). Another, web-tool ISMU has been developed by ICRISAT which provides GS analysis and methods but only to internal users. The above-mentioned tools were developed for a particular organism such as wheat or barley and thus may limit their use and exploration of GS in other crops. In order to combat this bottle neck, Tecle et al. (2014) developed a web-based tool called solGS, which can be used for GS studies in any organism. This tool predicts the GEBV of individuals using RR-BLUP. This tool has a graphical user interface and results are visualized in graphics. Results can also be downloaded in text format. Besides estimation of breeding values, solGS also performs correlation and heritability analysis of traits and calculates selection indices of individuals. The scheme is curated and implemented by several communities including *Solanaceae* Genomics Network, Cassavabase, Genome database for Rosaceae and Citrus genome database (Tecle et al., 2014). Apart from the above mentioned tools, GS analysis can also be done using any scripting language statistical software such as R or SAS. To conduct GS using R and SAS, scripts have been made available, although they are continuously evolving to enhance the prediction accuracy (Newell and Jannink, 2014; Charmet et al., 2020).

Genomic selection in *Solanaceae*

A comprehensive book has been published on GS in crop improvement and points out the wide use of GS in several crops including cereals, small grain crops, legumes, trees, and clonally propagated plants (Varshney et al., 2017). However, in case of *Solanaceae* crops, only few studies are reported suggesting that a lot of effort must be made for the improvement of these crops using advanced breeding techniques. As per our understanding, among *Solanaceae* plants, tomato is the only crop where GS have been used for crop improvement, for traits such as disease resistance, fruit quality, fruit weight, soluble solid content, fruit size, flavour, metabolites content and yield (Duangjit et al., 2016; Bautista et al., 2016; Yammmaoto et al., 2016; 2017; Liabeuf et al., 2018).

Bautista et al. (2016) conducted GS for six fruit related traits including—fruit weight, polar fruit diameter, equatorial fruit diameter, locule number, total soluble solid content and total yield per plant using select individuals in $F_{2:3}$ families. Three different approaches were used to make training and breeding populations from the F_2 generation. To predict breeding value, multiple linear regressions, ridge regression and Bayesian LASSO were used. Studies suggested that prediction accuracy varied among traits, environments, and models. Among three models used, Bayesian LASSO and ridge regression outperformed multiple linear regression. Among six traits studied, prediction accuracy was found highest for highly heritable traits. Among three different approaches for making training and validation populations, highest accuracy was found for multiple environment models. Complexity of traits also found to have an effect on prediction accuracy. Although, GS was suggested as a potential breeding approach in tomato, contradictorily, phenotypic selection was found to have higher genetic gain than GS in this study. A deep perusal of the study revealed that it suffered with several limitations such as small population size and lower marker density which might be confounding and may have resulted in the lower genetic gain of GS. Further, the effect of different parameters like training population size, marker density, and relatedness between the training and breeding populations on the GS prediction accuracy in tomato was discussed by Duangjit et al. (2016) based on GS performed for 45 traits including 35 metabolite and 10 quality traits respectively. In this study, a total 163 accessions representing three *Solanum* (*S. lycopersicum* L., *S. lycopersicum* var. *cerasiforme* and *S. pimpinellifolium*) sub species were used and genotyped with 7720 SNPs. The model RR-BLUP was used to make genomic predictions. To see the effect of training population size, different accession sample sizes were used as training and validation populations in the model. The highest prediction accuracy was observed when the training population was composed of the largest number of accessions in most cases. However, for some traits like soluble solid content, the highest prediction accuracy was observed in smaller samples from the training population. Studies also suggested that prediction accuracy is directly proportional to marker density. Outperformance was observed when the training and validation sets were more related and belonged to the same subpopulation.

Different statistical models have also been tested for prediction accuracy in tomato. GS for soluble solid content and fruit weight using six different GS models suggested that different models may be suitable for GS for different traits (Yammamoto et al., 2017). In this study, GS were conducted in 91 F_1 tomato varieties derived from crosses between four parents in four cross combinations. Six models including four linear models (GBLUP, Bayesian LASSO, weighted Bayesian shrinkage regression, and Bayes C) and two non-linear models (reproducing kernel Hilbert space regression and random forest) were tested and found that linear models are suitable (in terms of prediction accuracy) for soluble solid content, while non-linear models are suitable for fruit weight; suggested that non-linear factors contribute to fruit weight. Further, study proved that prediction accuracy is directly proportional to the heritability of the trait. Simulations were performed to test the breeding design along with phenotype prediction and have proved to enhance the improvement of crop production and quality substantially when applied for yield and flavour (Yammamoto et al., 2016). GS has also been conducted for disease resistance in tomato (Liabeuf et al., 2018). GS for bacterial spot disease resistance have been conducted using 109 tomato families derived from six highly resistant advanced breeding lines. In this study, 109 families were considered as training populations; and 51 inbred lines and 44 hybrids derived from 109 families were tested for validation purposes. It has been observed that prediction accuracy is higher in inbred populations compared to that in hybrids. Further, GS was conducted in two ways—one using only resistance associated markers and another using random markers distributed throughout the genome. The prediction accuracy was found higher when associated markers were used; however, marker density has a direct effect on prediction accuracy when random markers are used (Liabeuf et al., 2018).

Perspectives

With the advancement in genomics, lots of tools including high throughput marker systems and genome sequences are available in *Solanaceae* crops particularly major vegetables such as tomato, *Capsicum*, eggplant and potato.

These genomic tools are proving to be important assets for the *Solanaceae* breeding program. Most of the agricultural traits are complex in nature and controlled by a large number of minor effect genes/QTLs which are affected by the environmental conditions. In such situations, conventional breeding or MAS for one or two genes hardly prove to be useful for trait improvement. Therefore, GS, an advanced breeding technique of the present era, is being used for the improvement of several crops such as wheat, rice, legumes and maize for a number of traits. In case of *Solanaceae* crops, also with high density marker systems, whole genome sequences, availability of diverse germplasms, GS will prove to be an effective technique for crop improvement. Although, till date, only a few efforts have been made in this direction mainly in tomatoes; however, GS has proved more successful over conventional breeding for this crop. Other *Solanaceae* crops are being ignored till date which could be due to the unavailability of a proper protocol, databases/ tools for analysis. Crop breeding programs need to focus more on efforts in developing resources and models which can be used for GS to facilitate major breeding strategies for the improvement of *Solanaceae* crops.

Acknowledgements

Thanks are due to the University of Potential for Excellence II grant to Jawaharlal Nehru University from the University Grants Commission, India, and Council of Scientific and Industrial research concerning project (MLP0201). Additional support was received from the Science and Engineering Board, Ministry of Science and Technology, Government of India for ECR award given to VJ. Further, VJ and VG acknowledge the DST-INSPIRE faculty awards received from the Department of Science and Technology, Ministry of Science and Technology, Government of India. CSIR-IHBT publication number-4657.

References

Aflitos, S., 100 Tomato Genome Sequencing Consortium, Schijlen, E., de Jong, H., de Ridder, D. et al. 2014. Exploring genetic variation in the tomato (*Solanum section lycopersicon*) clade by whole-genome sequencing. *Plant J.* 80: 136–48.

Akdemir, D., Sanchez, J. I. and Jannink, J. L. 2015. Optimization of genomic selection training populations with a genetic algorithm. *Genet. Sel. Evol.* 47: 38.

Akdemir, D. and Isidro-Sanchez, J. 2019. Design of training populations for selective phenotyping in genomic prediction. *Sci. Rep.* 9: 1446.

Bagga, S., Lucero, Y., Apodaca, K. et al. 2019. Chile (*Capsicum annuum*) plants transformed with the *RB* gene from *Solanum bulbocastanum* are resistant to *Phytophthora capsici*. *Plos One* 14: e0223213.

Barchi, L., Lefebvre, V., Sage-Palloix, A. M. et al. 2009. QTL analysis of plant development and fruit traits in pepper and performance of selective phenotyping. *Theor. Appl. Genet.* 118: 1157–71.

Barchi, L., Lanteri, S., Portis, E. et al. 2011. Identification of SNP and SSR markers in eggplant using RAD tag sequencing. *BMC Genomics* 12: 304.

Bautistaa, A., Lobato-Ortiza, R., Garcia-Zavalaa, J. J. et al. 2016. Implications of genomic selection for obtaining F2:3 families of tomato. *Sci. Hor*. 207: 7–13.

Ben Chaim, A., Borovsky, E., Rao, G. U. et al. 2003a. *fs3.1*: a major fruit shape QTL conserved in *Capsicum*. *Genome* 46: 1–9.

Ben Chaim, A., Borovsky, E., De Jong, W. et al. 2003b. Linkage of the *A* locus for the presence of anthocyanin and *fs10.1*, a major fruit-shape QTL in pepper. *Theor. Appl. Genet.* 106: 889–894.

Ben-Chaim, A., Borovsky, Y., Falise, M. et al. 2006. QTL analysis for capsaicinoid content in *Capsicum*. *Theor. Appl. Genet.* 113: 1481–1490.

Bennewitz, S., Bergauu, N. and Tissier, A. 2018. QTL mapping of the shape of type VI glandular trichomes in tomato. *Front. Plant Sci.* 9: 1421.

Brand, A., Borovsky, Y., Hill, T. et al. 2014. *CaGLK2* regulates natural variation of chlorophyll content and fruit color in pepper fruit. *Theor. Appl. Genet.* 127: 2139–2148.

Brekke, T. D., Stroud, J. A., Shaw, D. S. et al. 2019. QTL mapping in salad tomatoes. *Euphytica* 215: 115.

Brouwer, D. J., Jones, E. S. and St Clair, D. A. 2004. QTL analysis of quantitative resistance to *Phytophthora infestans* (late blight) in tomato and comparisons with potato. *Genome* 47: 475–492.

Calus, M. P. L., Meuwissen, T. H. E., de Roos, A. P. W. et al. 2008. Accuracy of genomic selection using different methods to define haplotypes. *Genetics* 178: 553–561.

Capel, C., Yuste-Lisbona, F. J., López-Casado, G. et al. 2017. QTL mapping of fruit mineral contents provides new chances for molecular breeding of tomato nutritional traits. *Theor. Appl. Genet.* 130: 903–913.

Causse, M., Saliba-Colombani, V., Lecomte, L. et al. 2002. QTL analysis of fruit quality in fresh market tomato: a few chromosome regions control the variation of sensory and instrumental traits. *J. Exp. Bot.* 53: 2089–2098.

Celik, I., Gurbuz, N., Uncu, A. T. et al. 2017. Genome-wide SNP discovery and QTL mapping for fruit quality traits in inbred backcross lines (IBLs) of *Solanum pimpinellifolium* using genotyping by sequencing. *BMC Genomics* 18: 1.

Charmet, G., Tran, L. G., Auzanneau, J., Rincent, R. and Bouchet, S et al. 2020. BWGS: A R package for genomic selection and its application to a wheat breeding programme. *PLoS ONE* 15: e0222733.

Cheng, J., Qin, C., Tang, X. et al. 2016. Development of a SNP array and its application to genetic mapping and diversity assessment in pepper (*Capsicum* spp.). *Sci. Rep.* 6: 33293.

Collins, A., Milbourne, D., Ramsay, L. et al. 1999. QTL for field resistance to late blight in potato are strongly correlated with maturity and vigour. *Mol. Breed.* 5: 387–398.

Crossa, J., Perez-Rodriuez, P., Cuevas, J. et al. 2017. Genomic selection in plant breeding: methods, models, and perspectives. *Trends Plant Sci.* 22: 961–975.

De Jong, W. S., Eannetta, N. T., De Jong, D. M. et al. 2004. Candidate gene analysis of anthocyanin pigmentation loci in the *Solanaceae*. *Theor. Appl. Genet.* 108: 423–432.

de los Campos, G., Hickey, J., Pong-Wong, R. et al. 2013. Whole-genome regression and prediction methods applied to plant and animal breeding. *Genetics* 193: 327–345.

Djian-Caporalino, C., Pijarowski, L., Fazari, A. et al. 2001. High-resolution genetic mapping of the pepper (*Capsicum annuum* L.) resistance loci Me_3 and Me_4 conferring heat-stable resistance to root-knot nematodes (*Meloidogyne* spp.). *Theor. Appl. Genet.* 103: 592–600.

Doganlar, S., Frary, A., Daunay, M. C. et al. 2002b. Conservation of gene function in the *Solanaceae* as revealed by comparative mapping of domestication traits in eggplant. *Genetics* 161: 1713–1726.

Duangjit, J., Causse, M. and Sauvage, C. 2016. Efficiency of genomic selection for tomato fruit quality. *Mol. Breed.* 36: 29.

El Mohtar, C. A., Atamian, H. S., Dagher, R. B. et al. 2007. Marker-assisted selection of tomato genotypes with the *I-2* gene for resistance to *Fusarium oxysporum* f. sp. *lycopersici* race. *Plant Dis.* 91: 758–762.

FAO. 2013. *FAOSTAT database* http://faostat.fao.org.

Fentik, D. A. 2017. Review on genetics and breeding of tomato (*Lycopersicon esculentum* Mill). *Adv. Crop Sci. Tech.* 5: 306.

Frary, A., Nesbitt, T. C., Grandillo, S. et al. 2000. *fw2.2*: a quantitative trait locus key to the evolution of tomato fruit size. *Science* 289: 85–88.

Frary, A., Doganlar, S., Daunay, M. C. et al. 2003. QTL analysis of morphological traits in eggplant and implications for conservation of gene function during evolution of Solanaceous species. *Theor. Appl. Genet.* 107: 359–370.

Fulladolsa, A. C., Navarro, F. M., Kota, R. et al. 2015. Application of marker-assisted selection for Potato Virus Y resistance in the University of Wisconsin potato breeding program. *Am. J. Potato Res.* 92: 444.

Fulop, D., Ranjan, A., Ofner, I. et al. 2016. A new advanced backcross tomato population enables high resolution leaf QTL mapping and gene identification. *Gene Genome Genet.* 6: 3169–3184.

Ganal, M. W., Bonierbale, M. W., Roeder, M. S. et al. 1991. Genetic and physical mapping of the patatin genes in potato and tomato. *Mol. Genome Genet.* 225: 501–509.

Ghislain, M., Trognitz, B., Herrera, M. D. et al. 2001. Genetic loci associated with field resistance to late blight in offspring of *Solanum phureja* and *S. tuberosum* grown under short-day conditions. *Theor. Appl. Genet.* 103: 433–442.

Grandillo, S., Ku, H. M. and Tanksley, S. D. 1999. Identifying the loci responsible for natural variation in fruit size and shape in tomato. *Theor. Appl. Genet.* 99: 978–987.

Grube, R. C., Radwanski, E. R. and Jahn, M. K. 2000. Comparative mapping of disease resistance within the *Solanaceae*. *Genetics* 155: 873–887.

Gurbuz, N., Uluisik, S., Frary, A. et al. 2018. Health benefits and bioactive compounds of eggplant. *Food Chem.* 268: 602–610.

Han, K., Jeong, H., Yang, H. et al. 2016. An ultra-high-density bin map facilitates high-throughput QTL mapping of horticultural traits in pepper (*Capsicum annuum*). *DNA Res.* 23: 81–91.

Hanson, P., Lu, S. F., Wang, J. F. et al. 2016. Conventional and molecular marker-assisted selection and pyramiding of genes for multiple disease resistance in tomato. *Sci. Hor.* 201: 346–354.

Hara-Skrzypiec, A., Sliwka, J., Jakuczun, H. et al. 2018. QTL for tuber morphology traits in diploid potato. *J. Appl. Genet.* 59: 123–132.

Hirakawa, H., Shirasawa, K., Miyatake, K. et al. 2014. Draft genome sequence of eggplant (*Solanum melongena* L.): the representative solanum species indigenous to the old world. *DNA Res.* 21: 649–660.

Huang, C. C., Cui, Y. Y., Weng, C. R. et al. 2000. Development of diagnostic PCR markers closely linked to the tomato powdery mildew resistance gene *Ol-1* on chromosome 6 of tomato. *Theor. Appl. Genet.* 101: 918–924.

Isidro, J., Jannink, J. L., Akdemir, D. et al. 2015. Training set optimization under population structure in genomic selection. *Theor. Appl. Genet.* 128: 145–158.

Jaiswal, V., Rawoof, A., Dubey, M. et al. 2019. Development and characterization of non-coding RNA based simple sequence repeat markers in Capsicum species. *Genomics* 7543: 30362.

Jeong, H. S., Jang, S., Han, K. et al. 2015. Marker-assisted backcross breeding for development of pepper varieties (*Capsicum annuum*) containing capsinoids. *Mol. Breed.* 35: 226.

Kim, S., Park, M., Yeom, S. I. et al. 2014 Genome sequence of the hot pepper provides insights into the evolution of pungency in *Capsicum* species. *Nat. Genet.* 46: 270–278.

Kimbara, J., Ohyama, A., Chikano, H. et al. 2018. QTL mapping of fruit nutritional and flavor components in tomato (*Solanum lycopersicum*) using genome-wide SSR markers and recombinant inbred lines (RILs) from an intra-specific cross. *Euphytica* 214: 210.

Kumar, A., Tiwari, K. L., Datta, D. et al. 2014. Marker assisted gene pyramiding for enhanced Tomato leaf curl virus disease resistance in tomato cultivars. *Biol. Plant.* 58: 792–797.

Kumar, A., Jindal, S. K., Dhaliwal, M. S. et al. 2019. Gene pyramiding for elite tomato genotypes against ToLCV (*Begomovirus* spp.), late blight (*Phytophthora infestans*) and RKN (*Meloidogyne* spp.) for northern India farmers. *Physiol. Mol. Biol. Plants* 25: 1197–1209.

Lefebvre, V., Pflieger, S., Thabuis, A. et al. 2002. Towards the saturation of the pepper linkage map by alignment of three intraspecific maps including known-function genes. *Genome* 45: 839–854.

Li, J., Wang, Y., Wen, G. et al. 2019. Mapping QTL underlying tuber starch content and plant maturity in tetraploid potato. *Crop J.* 7: 261–271.

Li, X., Xu, J., Duan, S. et al. 2018. Mapping and QTL analysis of early-maturity traits in tetraploid potato (*Solanum tuberosum* L.). *Int. J. Mol. Sci.* 19: 3065.

Liabeuf, D., Sim, S. C. and Francis, D. M. 2018. Comparison of marker-based genomic estimated breeding values and phenotypic evaluation for selection of bacterial spot resistance in tomato. *Phytopathol.* 108: 392–401.

Liang, Z., Gupta, S. K., Yeh, C. T. et al. 2018. Phenotypic data from inbred parents can improve genomic prediction in pearl millet hybrids. *Gene Genome Genet.* 8: 2513–2522.

Liu, J., Zheng, Z., Zhou, X. et al. 2015. Improving the resistance of eggplant (*Solanum melongena*) to *Verticillium* wilt using wild species *Solanum linnaeanum. Euphytica* 201: 463–469.

Livingstone, K. D., Lackney, V. K., Blauth, J. R. et al. 1999. Genome mapping in *Capsicum* and the evolution of genome structure in the *Solanaceae*. *Genetics* 152: 1183–1202.

Massa, A. N., Manrique-Carpintero, N. C., Coombs, J. J. et al. 2015. Genetic linkage mapping of economically important traits in cultivated tetraploid potato (*Solanum tuberosum* L.). *Gene Genome Genet.* 5: 2357–2364.

Massa, A. N., Manrique-Carpintero, N. C., Coombs, J. et al. 2018. Linkage analysis and QTL mapping in a tetraploid russet mapping population of potato. *BMC Genet.* 1: 87.

Meijer, D., Viquez-Zamora, M., vanEck, H. J. et al. 2018. QTL mapping in diploid potato by using selfed progenies of the cross *S. tuberosum* × *S. chacoense. Euphytica* 214: 121.

Mimura, Y., Minamiyama, Y., Sano, H. et al. 2010. Mapping for axillary shooting, flowering date, primary axis length, and number of leaves in pepper (*Capsicum annuum*). *J. JPN. Soc. Hortic. Sci.* 79: 56–63.

Miura, K., Shiba, H., Ohta, M. et al. 2012. *SlICE1* encoding a MYC-type transcription factor controls cold tolerance in tomato, *Solanum lycopersicum*. *Plant Biotechnol.* 29: 253–260.

Monforte, A., Friedman, E., Zamir, D. et al. 2001. Comparison of a set of alleleic QTL-NILs for chromosome 4 of tomato: deductions about natural variation and implications for germplasm collection. *Theor. Appl. Genet.* 102: 572–590.

Nadakuduti, S. S., Holdsworth, W. L., Klein, C. L. et al. 2014. *KNOX* genes influence a gradient of fruit chloroplast development through regulation of *GOLDEN2-LIKE* expression in tomato. *Plant J.* 78: 1022–1033.

Newell, M. A. and Jannink, J. L. 2014. Genomic selection in plant breeding. *In*: Delphine Fleury and Ryan Whitford (eds.). *Crop Breeding: Methods and Protocols, Methods in Molecular Biology*. vol. 1145, Springer Science, Business Media, New York.

Nguyen, C. V., Vrebalov, J. T., Gapper, N. E. et al. 2014. Tomato *GOLDEN2-LIKE* transcription factors reveal molecular gradients that function during development and ripening. *Plant Cell* 26: 585–601.
Nunome, T., Ishiguro, K., Yoshida, T. et al. 2001. Mapping of fruit shape and color development traits in eggplant (*Solanum melongena* L.) based on RAPD and AFLP markers. *Breed Sci.* 51: 19–26.
Olmstead, R. G., Bohs, L., Migid, H. A., Santiago-Valentin, E., Garcia, V. F. and Collier, S. M. 2008. A molecular phylogeny of the Solanaceae. *Taxon* 57: 1159–1181.
Ozkaynak, E., Devran, Z., Kahveci, E. et al. 2014. Pyramiding multiple genes for resistance to *PVY, TSWV* and *PMMOV* in pepper using molecular markers. *Europ. J. Hort. Sci.* 79: S233–239.
Pan, Y., Bradley, G., Pyke, K. et al. 2013. Network inference analysis identifies an APRR2-Like gene linked to pigment accumulation in tomato and pepper fruits. *Plant Physiol.* 161: 1476–1485.
Panthee, D. R., Piotrowski, A. and Ibrahem, R. 2017. Mapping Quantitative Trait Loci (QTL) for resistance to late blight in tomato. *Int. J. Mol. Sci.* 18: 1589.
Portis, E., Barchi, L., Toppino, L. et al. 2014. QTL mapping in eggplant reveals clusters of yield-related loci and orthology with the tomato genome. *Plos One* 9: e89499.
Potato Genome Sequencing Consortium, Xu, X., Pan, S., Cheng, S. Z. et al. 2011. Genome sequence and analysis of the tuber crop potato. *Nature* 475: 189–195.
Powell, A. L., Nguyen, C. V., Hill, T. et al. 2012. Uniform ripening encodes a *Golden 2-like* transcription factor regulating tomato fruit chloroplast development. *Science* 336: 1711–1715.
Prasanna, H. C., Sinha, D. P., Rai, G. K. et al. 2015. Pyramiding *Ty-2* and *Ty-3* genes for resistance to monopartite and bipartite tomato leaf curl viruses of India. *Plant Pathol.* 64: 256–264.
Qin, C., Yu, C., Shen, Y. et al. 2014. Whole-genome sequencing of cultivated and wild peppers provides insights into *Capsicum* domestication and specialization. *Proc. Natl. Acad. Sci. USA* 111: 5135–40.
Roorkiwal, M., Rathore, A., Das, R. R. et al. 2016. Genome-enabled prediction models for yield related traits in chickpea. *Fron. Plant Sci.* 7: 1666.
Ruffel, S., Gallois, J. L., Moury, B. et al. 2006. Simultaneous mutations in translation initiation factors *eIF4E* and *eIF*(iso)*4E* are required to prevent pepper veinal mottle virus infection of pepper. *J. Gen. Virol.* 87: 2089–98.
Saatchi, M., Miraei-Ashtiani, S., Javaremi, A. N. et al. 2010. The impact of information quantity and strength of relationship between training set and validation set on accuracy of genomic estimated breeding values. *Afr. J. Biotechnol.* 9.
Santa, J. D., Berdugo-Cely, J., Cely-Pardo, L. et al. 2018. QTL analysis reveals quantitative resistant loci for *Phytophthora infestans* and *Tecia solanivora* in tetraploid potato (*Solanum tuberosum* L.). *Plos One* 13: e0199716.
Särkinen, T., Bohs, L., Olmstead, R. G. et al. 2013. A phylogenetic framework for evolutionary study of the nightshades (Solanaceae): a dated 1000-tip tree. *BMC Evol. Biol.* 13: 214.
Sim, S. C., Van Deynze, A., Stoffel, K. et al. 2012. High-density SNP genotyping of tomato (*Solanum lycopersicum* L.) reveals patterns of genetic variation due to breeding. *Plos One* 7: e45520.
Sweeney, D. W., Sun, J., Taagen, E. et al. 2019. Genomic selection in wheat. pp. 273–302. *In: Applications of Genetic and Genomic Research in Cereals.* Woodhead Publishing Series in Food Science, Technology and Nutrition.
Tan, M. Y. A., Hutten, R. C. B., Visser, R. G. F. et al. 2010. The effect of pyramiding *Phytophthora infestans* resistance *genes RPi-mcd1* and *RPi-ber* in potato. *Theor. Appl. Genet.* 121: 117–125.

Tanaka, Y., Yoneda, H., Hosokawa, M. et al. 2014. Application of marker-assisted selection in breeding of a new fresh pepper cultivar (*Capsicum annuum*) containing capsinoids, low-pungent capsaicinoid analogs. *Sci. Hort.* 165: 242–245.

Tecle, I. Y., Edwards, J. D., Menda, N. et al. 2014. solGS: a web-based tool for genomic selection. *BMC Bioinformatics* 15: 398.

Thabuis, A., Lefebvre, V., Bernard, G. et al. 2004. Phenotypic and molecular evaluation of a recurrent selection program for a polygenic resistance to *Phytophthora capsici* in pepper. *Theor. Appl. Genet.* 109: 342–351.

Thorup, T., Tanyolac, B., Livingstone, K. et al. 2000. Candidate gene analysis of organ pigmentation loci in the *Solanaceae*. *Proc. Nat. Acad. Sci. USA* 97: 11192–11197.

Tomato-Genome-Consortium. 2012. The tomato genome sequence provides insights into fleshy fruit evolution. *Nature* 485: 635–641.

Tpppinoo, L., Brachi, L., Scalzo, R. L. et al. 2016. Mapping quantitative trait loci affecting biochemical and morphological fruit properties in eggplant (*Solanum melongena* L.). *Fron. Plant. Sci.* 7: 256.

Tsaballa, A., Pasentsis, K. and Tsaftaris, A. S. 2012. The role of a *Gibberellin 20-oxidase* gene in fruit development in pepper (*Capsicum annuum*). *Plant Mol. Biol. Rep.* 30: 556–565.

Van der Knaap, E. and Tanksley, S. D. 2003. The making of a bell pepper-shaped tomato fruit: identification of loci controlling fruit morphology in Yellow Stuffer tomato. *Theor. Appl. Genet.* 107: 139–147.

Vanshney, R. K., Roorkiwal, M. and Sorrells, M. E. 2017. *Genomic Selection for Crop Improvement.* Springer International Publishing. XII, 258 DOI 10.1007/978-3-319-63170-7_1.

Villumsen, T. M., Janss, L. and Lund, M. S. 2009. The importance of haplotype length and heritability using genomic selection in dairy cattle. *J. Anim. Breed. Genet.* 126: 3–13.

Wang, X., Xua, Y., Huc, Z. et al. 2018. Genomic selection methods for crop improvement: Current status and prospects. *Crop J.* 6 : 330–340.

Wen, J., Jiang, F., Weng, Y. et al. 2019. Identification of heat-tolerance QTLs and high-temperature stress-responsive genes through conventional QTL mapping, QTL-seq and RNA-seq in tomato. *BMC Plant Biol.* 19: 398.

Wiggans, G. R., VanRaden, P. M., Bacheller, L. R. et al. 2010. Selection and management of DNA markers for use in genomic evaluation. *J. Dairy Sci.* 93: 2287–2292.

Wu, J., Liu, S., He, Y. et al. 2012b. Genome-wide analysis of *SAUR* gene family in *Solanaceae* species. *Gene* 509: 38–50.

Wu, J., Peng, Z., Liu, S. et al. 2012a. Genome-wide analysis of *Aux/IAA* gene family in *Solanaceae* species using tomato as a model. *Mol. Genome Genet.* 287: 295–311.

Xu, Y., Wang, X., Ding, X. et al. 2018. Genomic selection of agronomic traits in hybrid rice using an NCII population. *Rice* 11: 32.

Yamamoto, E., Matsunaga, A., Onogi, A. et al. 2016. A simulation-based breeding design that uses whole-genome prediction in tomato. *Sci. Rep.* 6: 19454.

Yamamoto, E., Matsunaga, H., Onogi, A. et al. 2017. Efficiency of genomic selection for breeding population design and phenotype prediction in tomato. *Heredity* 118: 202–209.

Yates, H. E., Frary, A., Doganla, S. et al. 2004. Comparative fine mapping of fruit quality QTLs on chromosome 4 introgressions derived from two wild tomato species. *Euphytica* 135: 283–296.

Zhang, J., Zhao, J., Xu, Y. et al. 2015. Genome-wide association mapping for tomato volatiles positively contributing to tomato flavour. *Front. Plant Sci.* 9: 1421.

Zhang, L. H., Mojtahedi, H., Kuang, H. et al. 2007. Root-knot nematode resistance introgressed from *Solanum bulbocastanum*. *Crop Sci.* 47: 2021–2026.

Zhang, X., Wang, G., Dong, T. et al. 2019. High-density genetic map construction and QTL mapping of first flower node in pepper (*Capsicum annuum* L.). *BMC Plant Biol.* 19: 167.

Zhao, Y., Liu, W., Xu, Y. P. et al. 2013. Genome-wide identification and functional analyses of calmodulin genes in Solanaceous species. *BMC Plant Biol*. 13: 70.

Zygier, S., Chaim, A. B., Efrati, A. et al. 2005. QTLs mapping for fruit size and shape in chromosomes 2 and 4 in pepper and a comparison of the pepper QTL map with that of tomato. *Theor. Appl. Genet.* 111: 437–445.

7

Genomic Selection in Oilseed Brassica: Potential, Prospects and Challenges

Bangkim Rajkumar,[#] *Garima*[#] and *Priya Panjabi**

ABSTRACT

Recent advances and cost reduction in the next-generation sequencing techniques have immensely contributed not only in developing genome wide high-density markers but also assisted cost-efficient, high-throughput genotyping in most crops. This has been a crucial driving force for implementing Genomic Selection (GS) studies in plants. The efficacy of selection using GS has been demonstrated in several crops, more recently, in oilseed Brassica. GS has been conducted in *B. napus* for traits including seed quality, yield related traits, plant height and blackleg resistance. High accuracies of prediction were achieved for most traits, especially for those with high heritability. Consistent with previous studies, prediction accuracies of low heritability traits in Brassica crops improved with an increase in the size of the training population, within-sub-population predictions and use of multi-trait GS models. Based on experiences gained from other plant crops, this chapter summarizes the strategies and crucial factors to be considered for achieving high GS based prediction efficacy in Brassica crops. Further, GS studies in Brassica crops have been discussed which are evidence for the immense potential of GS based breeding strategies.

Department of Botany, University of Delhi, Delhi 110007.
[#] Both authors contributed equally.
* Corresponding author: ppanjabi09@gmail.com

Introduction

Brassica genus includes many economically important vegetable and oilseed crops (Rakow, 2004; Snowdon, 2007). The six most common cultivated Brassica species include three diploid species (*B. nigra*, *B. oleracea* and *B. rapa*) and three allotetraploid species (*B. juncea*, *B. napus* and *B. carinata*), whose genetic relationships are explained as the 'U triangle' (U and N, 1935). Oilseed Brassicas (*B. napus*, *B. juncea* and *B. rapa*) constitute the second most important oilseed crop, next only to soyabean (USDA, 2020), with a global production of 68.2 million metric tonnes. India is the fourth largest producer of oilseed Brassica with a total production of 7.7 million tonnes (11.28% of the global production).

Molecular breeding approaches in oilseed Brassica crops aim at increasing the crop yield to maintain adequate supply for the ever-increasing human population given the constraints of limited land for growth, unfavourable climatic conditions and biotic and abiotic stress. Understanding the genetic architecture of complex traits is crucial to improve crop productivity since most agronomic traits (including morphology, yield, seed quality traits) are influenced by several genes and also by the environment and genotype-by-environment (G $\times$ E) interactions. Loci governing these complex traits have been identified in oilseed Brassica crops both through family based (linkage mapping) and diverse population based (linkage disequilibrium; LD mapping) approaches (Paritosh et al., 2013; Yang et al., 2017; Akhatar et al., 2020; Wang et al., 2020; Xuan et al., 2020; Wang et al., 2021). QTL mapping approaches have led to the identification of loci related to various agronomically important traits in *B. napus* including seed oil content and quality (Delourme et al., 2006; Qiu et al., 2006; Sun et al., 2012; Yang et al., 2012), siliqua related traits (Wang et al., 2016), pod number and seed number per pod (Shi et al., 2015). Similarly, in *B. juncea* a number of QTL underlying yield-associated traits (Ramchiary et al., 2007a; Yadava et al., 2012), oleic acid content (Sharma et al., 2002), erucic acid (Gupta et al., 2004; Rout et al., 2018), glucosinolates (Mahmood et al., 2003; Ramchiary et al., 2007b; Yan et al., 2020), white rust resistance (Panjabi-Massand et al., 2010; Arora et al., 2019; Bhayana et al., 2020) and resistance to blackleg disease (Christianson et al., 2006) have been identified.

More recently, genome-wide association studies (GWAS) in Brassica crop plants have been successfully implemented to identify loci related to various traits. For example, in *B. napus*, GWAS based approaches have led to identification of loci associated with traits such as seed oil content (Liu et al., 2016; Wu et al., 2016a; Fu et al., 2017; Xiao et al., 2019; Wang et al., 2021), fatty acid content (Qu et al., 2017; Tang et al., 2019), plant height (Sun et al., 2016a), branch number (He et al., 2017), branch angle (Sun et al.,

2016b) and flowering time (Xu et al., 2016). GWAS studies for abiotic and biotic stress related traits such as salt tolerance (Wassan et al., 2021), drought stress (Khanzada et al., 2020), freezing tolerance (Chao et al., 2021), clubroot resistance (Dakouri et al., 2021), resistance to *Sclerotinia* (Wu et al., 2016b) and blackleg resistance (Rahman et al., 2016) have also been conducted. GWAS studies have been reported in *B. juncea* for yield related traits (Gupta et al., 2021), pod shattering (Kaur et al., 2020) and seed quality traits (Akhatar et al., 2020).

QTL linked markers, identified through linkage based mapping and more recently through GWAS have led to the replacement of the traditional, phenotype based selection (PS) with marker assisted selections (MAS), thereby facilitating crop breeding initiatives. MAS, which relies on indirect selections based on markers linked with the trait loci, is effective only for traits that are governed by few, large-effect QTL. However, when a trait is controlled by several QTL with small effects, MAS is less efficient, sometimes even inferior to traditional phenotypic selection methods (Zhao et al., 2014). Numerous QTL mapping endeavours highlight the fact that most agronomic traits are governed by several small-effect loci (reviewed by Bernardo et al., 2008), whose individual contribution to the total genetic variation is limited (hence not detected through QTL mapping), but the cumulative effect might be high, sometimes even higher than the few large-effect QTL (Manolio et al., 2009). To overcome the limitations of MAS, an alternative approach of genomic selection (GS) was proposed by Meuwissen et al. (2001).

Genomic selection

GS is an extension of MAS, wherein instead of a few markers linked with the identified, large-effect QTL, genome-wide, high-density markers are used to predict the phenotypic performance of an individual. GS methodology utilizes two populations: training/reference population (TP) and a breeding population (BP). The training population comprises of individuals that are both genotyped (using genome-wide markers) and phenotyped (suitably across different environments). Use of genome-wide markers increases the possibility that all or most of the QTL contributing to the phenotype would be in LD with at least one marker. Cumulative effects of each assayed marker on the phenotype are estimated to train and develop a statistical model. This prediction model can then be used to estimate genomic estimated breeding values (GEBVs) for the selection of high performing, non-phenotyped individuals from a target BP. Individuals are thus selected based purely on their GEBVs (Meuwissen et al., 2001).

The statistical model developed is tested for its prediction efficacy using a cross-validation method, prior to using it for selection. In this technique, for example, the phenotypic and the genotypic data of the entire reference population is compiled and this dataset is then randomly split into k subsets containing information from an equal number of individuals. In each round of cross-validation, k–1 subsets are used as the TP, while the remaining subsets are used as the validating population (VP). The statistical model is generated based on the phenotypic and the genotypic data of the TP, and the predictive ability of the model is tested by estimating the GEBVs of the individuals of the VP, whose true breeding value (TBV) or phenotype is masked. The cross-validation is repeated k times and each time, a random set of individuals is part of the TP/VP. The accuracy of the model is based on the level of correlation between the true breeding value (TBV) and the predicted GEBVs (Habier et al., 2007).

Selections based on GEBVs not only shorten the breeding cycle, but also help save time and costs involved in extensive phenotyping thereby increasing the gains per unit time (Lorenzana and Bernardo, 2009; Heffner et al., 2010). Several GS studies have reported higher genetic gains as compared to PS or MAS based selections. For example, in common buckwheat (*Fagopyrum esculentum* Moench), GS led to a 20.9% increase in the selection index for yield related traits as compared to 15% by PS (Yabe et al., 2018). Similarly, GS was found to outperform PS in soft red winter wheat for various agronomic and yield related traits and an additional 10% selection gain was obtained when a combined PS and GS approach was used (Lozada et al., 2019).

Factors affecting prediction accuracies in GS

The prediction accuracy in GS depends on multiple factors. These include: (a) training population size and composition, (b) relatedness between the TP and BP, (c) marker density, (d) trait heritability and complexity, and (e) prediction models (Hayes et al., 2009; Hickey et al., 2014; Crossa et al., 2016; Schopp et al., 2017; Norman et al., 2018). These factors and their effects on the GS based prediction accuracy have been discussed below.

Training population design: size, composition and relatedness with test population

Both the composition (genotypes used) and size of the TP is detrimental for the success of GS. Genetic relatedness shared between TP and BP affect the GS prediction accuracies significantly and hence, the composition of the TP is decided based on the composition of the target BP (Edwards et al., 2019). GS is most effective in cases where the TP and BP are closely related (Cooper et al., 2014; Meuwissen et al., 2016). Hence, it is crucial to define the BP first and based on that the TP composition can be generated. The TP can be composed of individuals of a single or multiple bi-parental populations or accessions of a germplasm. Several studies have investigated the effect of TP composition on the prediction accuracies (Edwards et al., 2019; Lozada et al., 2019; Adeyemo et al., 2020). In one such extensive study by Riedelsheimer et al. (2013), five interconnected maize doubled haploid (DH) populations were used. Predictions made within full-sib families had high accuracies (0.59) even with a TP having 84 individuals. With the same TP size, predictions based on half-sib DH lines drastically reduced the accuracies (~ 42% decline). However, TPs composed of half-sib lines representing both parents of the VP, substantially increased the prediction accuracies, with no effect seen on inclusion of more crosses if the TP size was kept constant.

TP size is another important parameter to be considered for GS studies. Efforts to optimize the TP size focus to achieve desirable prediction accuracies with a smaller set of individuals so as to minimize both genotyping and phenotyping costs (Isidro et al., 2015). The optimum size of the TP is dependent on multiple factors including its relatedness with BP, level of genetic diversity within the TP, heritability of the trait (h^2), complexity of the trait and population structure (reviewed by Bassi et al., 2016). In general, increasing the TP size improves the prediction accuracy even for traits with low heritability, since the predictions are developed using larger representations of phenotypes and genotypes (Hickey et al., 2014; Zhang et al., 2017; Edwards et al., 2019; Liu et al., 2019). However, beyond a certain size, depending on factors specific to each study design, a further increase in TP size provides no additional advantage (Habier et al., 2007; Heffner et al., 2010; Zhao et al., 2012). For example, in a study using 988 F_6 advanced breeding lines in wheat, an increment in the TP size (from 49 to 934 individuals) initially increased the prediction accuracies until a plateau was reached at a TP size of ~ 700 (Cericola et al., 2017).

Increasing the TP size (several hundreds of individuals) can compensate for less relatedness shared between the TP and BP. Bassi et al. (2016) suggested that in wheat, prediction accuracies > 0.5 can be achieved with a TP size of at least 50 in case of full-sib populations, 100 individuals for TPs comprising of half-sibs and at least 1000 individuals in case of unrelated TPs.

Heritability of the trait

Heritability, the proportion of phenotypic variation that is due to genetic effects, is another significant factor contributing to GS prediction accuracy. In general, traits with high heritability show high GS prediction accuracies (Hayes et al., 2009; Lin et al., 2014; Fernandes et al., 2018). For low heritability traits, prediction accuracy has been shown to substantially increase by either using larger TPs or multi-trait GS models (Jia and Jannink, 2012; Combs and Bernardo, 2013; Wang et al., 2017; Mrode et al., 2019).

Additionally, in cases where the reference population exhibits population structure, prediction accuracy of low heritability traits can be enhanced by using within-population, instead of across-population predictions (Jan et al., 2016). In a population, several traits are highly correlated, although they might vary in their individual heritability. Multi-trait model based predictions take advantage of high correlations shared between a particular low heritability trait with an auxiliary higher heritability trait. Several studies have reported an increase in prediction accuracies of low heritability traits when multi-trait models were used (Jia and Jannik, 2012; Hayashi and Iwata, 2013; Guo et al., 2014a; Wang et al., 2017).

Population structure

Population structure is another crucial determinant of the prediction accuracies obtained. A population having a strong structure suggests that the constituent subpopulations differ significantly in their allele frequencies. Genomic predictions for traits that exhibit sub-population specific values would be biased in a population with high structure. Since, in such cases it would be difficult to differentiate the real marker-trait associations from those alleles that vary because of the structure and hence appear to be falsely associated with the trait (Isidro et al., 2015; Norman et al., 2018; Rio et al., 2018). In one of the several studies (Guo et al., 2014b; Zhang et al., 2017; Lyra et al., 2018) to evaluate the impact of population structure on prediction accuracies, Norman et al. (2018) utilized a population of 10,375 wheat lines that represented five subpopulations/clusters. The prediction accuracies based on 'all-cluster' analysis were similar to 'within-cluster' analysis for traits; glaucousness, and thousand kernel weight, while they were slightly higher for grain yield. Both analyses were similar in the respect that irrespective of the constitution (in terms of the number of subpopulations represented in each case), the TP was representative of the VP. This further strengthens the fact that more than the size and extent of diversity in the TP, the relatedness shared between the TP and BP are crucial for higher prediction accuracies.

Marker density

GS studies require well-spread markers that represent the whole genome. The number of markers for GS should be sufficient to ensure that each QTL is in strong LD with at least one of the markers, thereby ensuring that the prediction model is developed based on the phenotypic variance of most/all loci contributing to a trait (Meuwissen et al., 2001). High-density markers would ensure the fulfilment of this requirement, however, it would also lead to a substantial increase in the genotyping costs. Prediction accuracies in several studies were seen to increase initially with increasing marker density and then plateau after a certain density was reached (Habier et al., 2007; Zhao et al., 2012). In one such study, Liu et al. (2018) found that the prediction accuracies initially increased with an increase in the number of markers and then plateaued at a marker density of 1000 and 7000 for bi-parental and natural populations, respectively. Hence, even with a substantial decrease in marker density, prediction accuracies comparable to those with high density markers (37,803) were achieved in both populations. Similar results were observed in several other studies (Heffner et al., 2011; Cericola et al., 2017; Hao et al., 2019). These studies clearly demonstrate that the density of markers is dependent on the LD structure of the population. An optimum marker density would therefore vary across species and populations depending largely on their structure and extent of LD. High prediction accuracies can be achieved with fewer markers (in hundreds) in bi-parental populations with high LD than in multi-family (in thousands) or natural populations with considerably low LD structure (Zhao et al., 2012; Crossa et al., 2014; Liu et al., 2018).

Prediction models

For most complex traits, the total variance is composed of both additive and non-additive effects. GS prediction models developed initially, accounted for the additive gene effects only, while the non-additive effects (dominance and epistasis) were ignored as they were difficult to dissect. Models based on additive gene effects include: Ridge regression–best linear unbiased prediction (RR-BLUP), Genomic best linear unbiased prediction (GBLUP), Bayesian methods (BayesA and BayesB; Meuwissen et al., 2001) and machine learning (reviewed by Desta and Ortiz, 2014; Wang et al., 2018). Both GBLUP (VanRaden, 2008; Hayes et al., 2009) and RR-BLUP (Whittaker et al., 2000; Meuwissen et al., 2001) assume a common variance for all loci. Hence, both models are considered more suitable for polygenic traits which are controlled by a large number of small effect QTL as observed in maize (Riedelsheimer et al., 2013; Xu et al., 2017), wheat (Zhao et al., 2013; Wang et al., 2015; Haile et al., 2018) and canola (Würschum et al., 2014; Zou et al., 2016). Bayesian methods on the other hand, attribute different weightage to different loci and are able to capture large-effect QTL (Meuwissen, 2009). Predictive abilities of Bayesian methods are therefore higher than GBLUP for traits with few large-effect QTL and some small-effect loci (Wolc et al., 2016).

Omission of non-additive gene effects can result in biased prediction abilities, especially for traits in which they constitute a significant fraction of the total variance. Therefore, more recently, several variants of the GS models are available which incorporate non-additive gene effects. Examples of these include: adaptive mixed LASSO (least absolute shrinkage and selection operator; Wang et al., 2011a), EG-BLUP (extended genomic best linear unbiased prediction; Henderson, 1985), RKHS (reproducing kernel Hilbert space) and regression models (Gianola et al., 2006; Gianola and Van Kaam, 2008). Inclusion of epistasis and dominance gene effects have been shown to significantly increase the prediction ability in several studies (Dudley and Johnson, 2009; Denis and Bouvet, 2013; Jiang and Reif, 2015).

In most initial GS studies, the prediction models deployed were based on single environment or single trait data. Several multivariate models are now available to allow incorporation of multi-trait (MT) or multi-environment (ME) data. MT models are superior to single-trait models in their predictive ability, especially for low heritability traits. Multiple studies have shown an increase in the prediction accuracies of low heritability traits by using MT models that allow inclusion of phenotypic data of a highly correlated, high heritability auxiliary trait (Guo et al., 2014a; Wang et al., 2017). Additionally, MT models can also benefit predictive ability of traits that are otherwise difficult to phenotype at the early stages of plant development or are cost- and/or labour-intensive. Predictions for such traits can be done indirectly by incorporating phenotypic data of other auxiliary traits in the MT analysis that are easier/cost-effective to measure and are highly correlated with the trait of interest.

Also, many quantitative traits are affected by environmental factors and accounting for G × E interactions for such traits in the GS prediction models is crucial to facilitate selection of lines that perform similarly across environments and also in specific untested environments. Several GS models that allow multi-environment (ME) analysis have been developed (Burgueño et al., 2012; Jarquín et al., 2014) and demonstrated to significantly increase the prediction accuracies as compared to single-environment models (Heslot et al., 2014; Lopez-Cruz et al., 2015; Lado et al., 2016; Roorkiwal et al., 2018; Mageto et al., 2020).

Whole genome sequencing (WGS) of Brassica species

Owing to the cost-effective, high-throughput next generation sequencing (NGS) platforms available, multiple whole genome sequencing of Brassica genomes has been achieved. These sequencing endeavours provide a useful resource for mining of genome wide polymorphic markers that have accelerated efforts for establishing GS based breeding strategies in the crop plant. The first Brassica genome sequence published was of *B. rapa* (Wang et al., 2011b) followed by *B. oleracea* (Liu et al., 2014), *B. napus* (Chalhoub et al., 2014), *B. nigra* and *B. juncea* (Yang et al., 2016).

In case of *B. napus*, genomes of various other accessions have been sequenced after the first genome assembly (Chalhoub et al., 2014). These include cultivars Tapidor (Bayer et al., 2017), ZS11 (Sun et al., 2017; Chen et al., 2021), NY7 (Chen et al., 2021), Darmor-*bzh* (Rousseau-Gueutin et al.,

2020) and Express 617 (Lee et al., 2020). The most recent, much improved and highly contiguous version of genome assembly (921.5 Mb; scaffold N50 of 53.2 Mb), was *de novo* assembled using PacBio SMRT, Hi-C techniques (Chen et al., 2021) and contains 106, 059 predicted genes. Apart from this, a pan-genomic comparative analysis using *de novo* assembly of eight accessions of *B. napus* (Westar, Zheyou7, Tapidor, Quinta, No2127, Gangan, Shengli, ZS11) by Song et al. (2020), enabled identification of 77.2–149.6 Mb presence and absence variants (PAVs), a highly resourceful genetic resource for genomic studies.

In case of *B. juncea*, two *de novo* WGS have been achieved. The first draft genome (var. tumida) was published by Yang et al. (2016). The 955 Mb assembly (scaffold N50 of 1.5 Mb) with 80, 050 annotated genes was assembled using Illumina HiSeq™ shotgun, BioNano sequencing (optical mapping), and PacBio single molecule real time (SMRT) sequencing. More recently, an improved version of the genome (cv. Varuna) using long read SMRT sequencing, Illumina HiSEq and optical mapping was published by Paritosh et al. (2021). The 18 pseudo-chromosomes represent 840.2 Mb of the genome and a total of 101, 959 genes were predicted.

GS studies in oilseed Brassica crops

Given the promising results shown by genomic selection in cereal crops, application of GS in non-cereal crops is also gaining significance. Recently, several studies exploring the potential of GS in breeding of oilseed Brassica crops (specifically *B. napus*) have been published.

One of the earliest studies of genomic selection in *B. napus* was conducted by Würschum et al. (2014), to explore the potential of using GS in rapeseed breeding. The study used a DH population of 391 individuals (generated from nine different crosses), genotyped using 253 SNPs and phenotyped for six traits. For most traits, both models (RR-BLUP and BayesB) generated similar prediction accuracies, the highest (~ 0.8) being for plant height, a highly heritability and less complex trait. For grain yield and oil content, both traits with complex architecture, RR-BLUP performed better. Incorporation of family effects in predication models had no effect on prediction accuracy, despite significant differences in the trait mean values between families. Exclusion of markers around the previously identified QTL regions (Würschum, 2012) led to a small decrease in the prediction accuracy of all traits, with maximum decrease (~ 22%) seen for flowering time trait and lowest (~ 1.4%) for protein content. In case of flowering time, a relatively less complex trait, the identified QTLs explain most of the observed genetic variance and hence exclusion of these QTL regions significantly decreased the prediction accuracy.

A subsequent study by Jan et al. (2016), explored the potential of GS for predicting testcross performance that would enable effective pre-selection of promising parental combinations based purely on their genotype. A major aspect of the study was to decipher the effect of population structure on prediction accuracies. A population of 475 spring type canola pollinator lines were used to develop 950 hybrids that were phenotyped for seven traits across four countries. The pollinator lines, genotyped using 24,403 SNPs, comprised of three subpopulations/clusters (C1, C2 and C3). Predictions were performed in three independent ways: (a) across whole population, wherein predictions were performed and also validated using the whole population, (b) within-subpopulation, wherein predictions were performed and accuracy tested independently within each of the two subpopulations: C1 and C2 and (c) predictions made across the whole population and validated separately within C1 and C2. Across the whole population, prediction accuracy was highest (0.81) for seed oil content, probably owing to the high heritability and less complex architecture of the trait, and lowest for seedling emergence (0.29), the least heritable of the seven traits. Improvement in prediction accuracies were observed in low heritability trait (seedling emergence) in case of within-subpopulation predictions, in the larger of the two subpopulations having relatively narrow genetic diversity. In the third case, the prediction accuracies were negative or almost zero. For the high heritability traits, seed oil content and oil yield, no difference was observed in the across- and within-population predictions.

Similar results were observed in a subsequent study by Fikere et al. (2018), wherein, the prediction accuracies across subpopulations were much reduced in comparison to within-subpopulation predictions. These studies in *B. napus*, similar to observations in other crop plants (Guo et al., 2014b; Isidro et al., 2015), re-iterate that prior to performing GS, it is crucial to assess the structure of the populations being used. For populations with a strong structure, across-subpopulation predictions should be avoided owing to the limited relatedness shared between the TP and VP.

Zou et al. (2016) explored the effect of: (a) TP size variation, (b) marker density, and (c) the phenotype intensity (number of environments used for phenotyping) on prediction accuracies in *B. napus*. Consistent with previous studies of similar kind (Hickey et al., 2014; Jan et al., 2016; Schopp et al., 2017), high prediction accuracies could be achieved even with a small TP size in the case of bi-parental rapeseed populations. The study used a DH population of 180 individuals (Qiu et al., 2006), genotyped using 13, 678 SNPs (Zhang et al., 2016) and phenotyped for six traits across eleven environments. Genomic predictions were performed using four different models-GBLUP, RR-BLUP, BayesC (pi) and EG-BLUP. The prediction accuracy was found to increase with an increase in TP size. Cross-validations performed with phenotypic data from random subsets of the total environments (2 to 11, in increments of one, in case of oil content and 2–5 for protein content) revealed that high prediction accuracies (of 0.73 and 0.60) were reached using data from across only two environments for both traits. These predictions were marginally lower (3%–6%) as compared to that with a complete dataset (all environments). In the study, TP size was found to be the more crucial determinant of prediction accuracies than the number of environments used for phenotyping. An increase in marker density increased the prediction accuracies and reached a plateau at ~ 1000 markers. Also, a reduction in the markers (1,527 SNPs) achieved by the removal of redundant SNPs exhibiting strong LD, yielded prediction accuracies comparable with the complete marker set (13,678 SNPs). The study re-iterated the fact that in populations with highly conserved and extensive LD (which is the case in most breeding populations), even low-density markers, carefully selected considering the LD structure are sufficient to achieve high prediction accuracies. This would allow for a substantial reduction in genotyping costs without compromising on the prediction quality.

Similar observations regarding marker density were observed in an extensive study by Werner et al., 2018b, wherein two different approaches of marker reduction and its effect on the prediction accuracies were explored. The study utilized 203 *B. napus* inbred lines, genotyped using 24,338 SNPs and phenotyped for three traits. Several reduced marker subsets were generated using two reduction strategies: (a) haplotype based trait-independent 'tag-SNPs' wherein markers are selected based on the LD structure of the population (de Bakker et al., 2005) and (b) trait specific GWAS-assisted marker selection. An initial set of 9793 'tag-SNPs', sufficient to represent the whole genome was identified, from which several additional reduced subsets were generated. Of the two approaches, prediction accuracies using 'tag-SNP' reduced subsets (even < 1000 SNPs) were better especially for highly polygenic traits (oil content and plant height). In contrast, GWAS based strategy that does not rely on genome-wide coverage, performed poorly for such polygenic traits as it was not effective in capturing the small-QTL effects. Although, in case of the simple trait (glucosinolate content), GWAS based marker reduction approach did perform better.

In a subsequent study (Werner et al., 2018a), six different GS models, considering general and specific combining ability (GCA and SCA) were compared for predicting hybrid performance in *B. napus*. A total of 448 hybrid lines were genotyped using 13, 550 SNPs and phenotyped for seven traits across five locations. The models included three RR-BLUP based (GCA RR-BLUP, GCA RR-BLUP + *de novo* GWAS and GCA + SCA RR-BLUP) and three Bayesian models (GCA BayesB, GCA + SCA BayesB and GCA-BRR

+ SCA-BayesB). All the models gave comparable prediction values, for most traits. Inclusion of SCA did not impart any additional enhancement in the prediction values. Another model, GS + *de novo* GWAS method (Spindel et al., 2016) was also tested. This model allows for attributing fixed-effect to GWAS identified markers, thereby making them immune to the shrinkage factor and ensuring attribution of a stronger effect to the GWAS identified major QTL. Inclusion of GWAS based SNPs identified for two traits (flowering time and glucosinolate content) in the GS + *de novo* GWAS method, led to higher prediction accuracy (in comparison with other models) of one of the two traits (glucosinolate content).

The effect of inclusion of prior QTL information on genomic prediction accuracies, was also explored in a study by Fikere et al. (2018). Another major aspect of the study was to investigate the efficacy of GS based predictions for a trait like blackleg resistance in canola, wherein the previously identified QTLs explain only a fraction (< 30%) of the genetic variance and a lot remains undetected. For this, 532 canola lines were genotyped using 98,054 SNPs and phenotyped for three traits: emergence count, adult plant survival rate and internal infection of stem at maturity. Several of the blackleg resistance conferring QTL were identified from previous studies. Using the GBLUP model, the genetic variances were estimated for SNPs located within QTL regions and also separately for SNPs in the remaining genome. Interestingly, the genetic variance for survival and internal infection explained by the QTL regions was substantially lower (0.02 to 0.33) than the rest of the genome. These results signified the immense potential of GS in detecting several QTL present in the genomes that were otherwise not detected through the traditional QTL mapping owing to their minor individual contribution to the overall genetic variance. More significantly, the cumulative effect of these small-effect QTL identified through GS was much higher than the previously reported large-effect QTL. Use of prior QTL information was found to increase prediction accuracies for survival and internal infection using BayesRC.

Another study in *B. napus*, investigated the effect of inclusion of G × E interactions (in GBLUP models) for genomic predictions (Fikere et al., 2020). The study used 202 spring canola lines that were phenotyped for 22 agronomic traits including the blackleg resistance across several years at different locations including rain-fed and irrigated sites. For this, different environment factors (year, location and water conditions, i.e., rain-fed/irrigated) were added one at a time to generate four GBLUP model variants. An additional GBLUP model variant included all factors together. Incorporation of G × E interactions was found to increase the prediction accuracy (0.2% to 6%) across traits. Maximum improvements in prediction accuracies were observed when all the environment factors were included in the GBLUP model.

Conclusions

Genomic selection, a method involving whole-genome marker based selection of untested (non-phenotyped) individuals, has been shown to be much more efficient than MAS in breeding for complex traits in several crop plants. Recent studies have tested the feasibility and prospects of applying GS methods for improving selection of desired lines in oilseed Brassica, so far only in the species *B. napus*. Similar to studies in other crops, higher prediction accuracies were achieved for high heritability traits in *B. napus*. Within-subpopulation rather than whole-population based predictions yielded better results for low heritability traits (e.g., seedling emergence), with no improvement observed for traits with high heritability (e.g., plant height and seed oil content). Additionally, use of multi-trait models also increased the prediction efficacy of low heritability traits. Low density, genome-wide markers, carefully chosen based on the LD structure of the Brassica population were shown to have prediction capabilities similar to high density markers and would therefore facilitate reductions in genotyping costs. Use of appropriate GS statistical models and inclusion of fixed-marker effects for known QTL, marker correlations and environment effects in different models have shown encouraging results for the improvement of both simple and complex traits in Brassica. GS, thus has immense potential for facilitating Brassica crop improvement and breeding strategies.

Acknowledgments

We would like to thank UGC and CSIR, Government of India for providing research fellowship to Bangkim Rajkumar and Garima, respectively.

References

Adeyemo, E., Bajgain, P., Conley, E., Sallam, A. H., Anderson, J. A et al. 2020. Optimizing training population size and content to improve prediction accuracy of FHB-related traits in wheat. *Agronomy* 10(4): 543.

Akhatar, J., Singh, M. P., Sharma, A., Kaur, H., Kaur, N. et al. 2020. Association mapping of seed quality traits under varying conditions of nitrogen application in *Brassica juncea* L. Czern & Coss. *Frontiers in Genetics* 11: 744.

Arora, H., Padmaja, K. L., Paritosh, K., Mukhi, N., Tewari, A. K. et al. 2019. BjuWRR1, a CC-NB-LRR gene identified in *Brassica juncea*, confers resistance to white rust caused by *Albugo candida*. *Theoretical and Applied Genetics* 132(8): 2223–2236.

Bassi, F. M., Bentley, A. R., Charmet, G., Ortiz, R., Crossa, J. et al. 2016. Breeding schemes for the implementation of genomic selection in wheat (*Triticum* spp.). *Plant Science* 242: 23–36.

Bayer, P. E., Hurgobin, B., Golicz, A. A., Chan, C. K. K., Yuan, Y. et al. 2017. Assembly and comparison of two closely related *Brassica napus* genomes. *Plant Biotechnology Journal* 15(12): 1602–1610.

Bernardo, R. 2008. Molecular markers and selection for complex traits in plants: learning from the last 20 years. *Crop Science* 48(5): 1649–1664.
Bhayana, L., Paritosh, K., Arora, H., Yadava, S. K., Singh, P. et al. 2020. A mapped locus on LG A6 of *Brassica juncea* line Tumida conferring resistance to white rust contains a CNL type R gene. *Frontiers in Plant Science* 10: 1690.
Burgueño, J., de los Campos, G., Weigel, K. and Crossa, J. 2012. Genomic prediction of breeding values when modeling genotype x environment interaction using pedigree and dense molecular markers. *Crop Science* 52(2): 707–719.
Cericola, F., Jahoor, A., Orabi, J., Andersen, J. R., Janss, L. L. et al. 2017. Optimizing training population size and genotyping strategy for genomic prediction using association study results and pedigree information. A case of study in advanced wheat breeding lines. *PloS One* 12(1): e0169606.
Chalhoub, B., Denoeud, F., Liu, S., Parkin, I. A., Tang, H. et al. 2014. Early allopolyploid evolution in the post-Neolithic *Brassica napus* oilseed genome. *Science* 345(6199): 950–953.
Chao, W. S., Horvath, D. P., Stamm, M. J. and Anderson, J. V. et al. 2021. Genome-wide association mapping of freezing tolerance loci in canola (*Brassica napus* L.). *Agronomy* 11(2): 233.
Chen, X., Tong, C., Zhang, X., Song, A., Hu, M. et al. 2021. A high-quality *Brassica napus* genome reveals expansion of transposable elements, subgenome evolution and disease resistance. *Plant Biotechnology Journal* 19(3): 615–630.
Christianson, J. A., Rimmer, S. R., Good, A. G. and Lydiate, D. J. 2006. Mapping genes for resistance to *Leptosphaeria maculans* in *Brassica juncea*. *Genome* 49(1): 30–41.
Combs, E. and Bernardo, R. 2013. Accuracy of genomewide selection for different traits with constant population size, heritability, and number of markers. *The Plant Genome* 6: 1–7.
Cooper, M., Messina, C. D., Podlich, D., Totir, L. R., Baumgarten, A. et al. 2014. Predicting the future of plant breeding: complementing empirical evaluation with genetic prediction. *Crop and Pasture Science* 65(4): 311–336.
Crossa, J., Perez, P., Hickey, J., Burgueno, J., Ornella, L. et al. 2014. Genomic prediction in CIMMYT maize and wheat breeding programs. *Heredity* 112(1): 48–60.
Crossa, J., Jarquín, D., Franco, J., Pérez-Rodríguez, P., Burgueño, J. et al. 2016. Genomic prediction of gene bank wheat landraces. *G3: Genes, Genomes, Genetics* 6(7): 1819–1834.
Dakouri, A., Lamara, M., Karim, M. M., Wang, J., Chen, Q. et al. 2021. Identification of resistance loci against new pathotypes of *Plasmodiophora brassicae* in *Brassica napus* based on genome-wide association mapping. *Scientific Reports* 11(1): 1–11.
de Bakker, P. I., Yelensky, R., Pe'er, I., Gabriel, S. B., Daly, M. J. et al. 2005. Efficiency and power in genetic association studies. *Nature Genetics* 37(11): 1217–1223.
Delourme, R., Falentin, C., Huteau, V., Clouet, V., Horvais, R. et al. 2006. Genetic control of oil content in oilseed rape (*Brassica napus* L.). *Theoretical and Applied Genetics* 113(7): 1331–1345.
Denis, M. and Bouvet, J. M. 2013. Efficiency of genomic selection with models including dominance effect in the context of Eucalyptus breeding. *Tree Genetics & Genomes* 9(1): 37–51.
Desta, Z. A. and Ortiz, R. 2014. Genomic selection: Genome-wide prediction in plant improvement. *Trends in Plant Science* 19(9): 592–601.
Dudley, J. W. and Johnson, G. R. 2009. Epistatic models improve prediction of performance in corn. *Crop Science* 49(3): 763–770.
Edwards, S. M., Buntjer, J. B., Jackson, R., Bentley, A. R., Lage, J. et al. 2019. The effects of training population design on genomic prediction accuracy in wheat. *Theoretical and Applied Genetics* 132(7): 1943–1952.

Fernandes, S. B., Dias, K. O., Ferreira, D. F. and Brown, P. J. 2018. Efficiency of multi-trait, indirect, and trait-assisted genomic selection for improvement of biomass sorghum. *Theoretical and Applied Genetics* 131(3): 747–755.

Fikere, M., Barbulescu, D. M., Malmberg, M. M., Shi, F., Koh, J. C. et al. 2018. Genomic prediction using prior quantitative trait loci information reveals a large reservoir of underutilised blackleg resistance in diverse canola (*Brassica napus* L.) lines. *The Plant Genome* 11(2): 170100.

Fikere, M., Barbulescu, D. M., Malmberg, M. M., Maharjan, P., Salisbury, P. A. et al. 2020. Genomic prediction and genetic correlation of agronomic, blackleg disease, and seed quality traits in canola (*Brassica napus* L.). *Plants* 9(6): 719.

Fu, Y., Zhang, D., Gleeson, M., Zhang, Y., Lin, B. et al. 2017. Analysis of QTL for seed oil content in *Brassica napus* by association mapping and QTL mapping. *Euphytica* 213(1): 1–15.

Gianola, D., Fernando, R. L. and Stella, A. 2006. Genomic-assisted prediction of genetic value with semiparametric procedures. *Genetics* 173(3): 1761–1776.

Gianola, D. and Van Kaam, J. B. 2008. Reproducing kernel Hilbert spaces regression methods for genomic assisted prediction of quantitative traits. *Genetics* 178(4): 2289–2303.

Guo, G., Zhao, F., Wang, Y., Zhang, Y., Du, L. et al. 2014a. Comparison of single-trait and multiple-trait genomic prediction models. *BMC Genetics* 15(1): 1–7.

Guo, Z., Tucker, D. M., Basten, C. J., Gandhi, H., Ersoz, E. et al. 2014b. The impact of population structure on genomic prediction in stratified populations. *Theoretical and Applied Genetics* 127(3): 749–762.

Gupta, N., Gupta, M., Akhatar, J., Goyal, A., Kaur, R. et al. 2021. Association genetics of the parameters related to nitrogen use efficiency in *Brassica juncea* L. *Plant Molecular Biology* 105(1): 161–175.

Gupta, V., Mukhopadhyay, A., Arumugam, N., Sodhi, Y. S., Pental, D. et al. 2004. Molecular tagging of erucic acid trait in oilseed mustard (*Brassica juncea*) by QTL mapping and single nucleotide polymorphisms in FAE1 gene. *Theoretical and Applied Genetics* 108(4): 743–749.

Habier, D., Fernando, R. L. and Dekkers, J. C. 2007. The impact of genetic relationship information on genome-assisted breeding values. *Genetics* 177(4): 2389–2397.

Haile, J. K., N'Diaye, A., Clarke, F., Clarke, J., Knox, R. et al. 2018. Genomic selection for grain yield and quality traits in durum wheat. *Molecular Breeding* 38(6): 1–18.

Hao, Y., Wang, H., Yang, X., Zhang, H., He, C. et al. 2019. Genomic prediction using existing historical data contributing to selection in biparental populations: a study of kernel oil in maize. *The Plant Genome* 12(1): 180025.

Hayashi, T. and Iwata, H. 2013. A Bayesian method and its variational approximation for prediction of genomic breeding values in multiple traits. *BMC Bioinformatics* 14(1): 1–14.

Hayes, B. J., Visscher, P. M. and Goddard, M. E. 2009. Increased accuracy of artificial selection by using the realized relationship matrix. *Genetics Research* 91(1): 47–60.

He, Y., Wu, D., Wei, D., Fu, Y., Cui, Y. et al. 2017. GWAS, QTL mapping and gene expression analyses in *Brassica napus* reveal genetic control of branching morphogenesis. *Scientific Reports* 7(1): 1–9.

Heffner, E. L., Lorenz, A. J., Jannink, J. L. and Sorrells, M. E. 2010. Plant breeding with genomic selection: Gain per unit time and cost. *Crop Science* 50(5): 1681–1690.

Heffner, E. L., Jannink, J. L., Iwata, H., Souza, E., Sorrells, M. E. et al. 2011. Genomic selection accuracy for grain quality traits in biparental wheat populations. *Crop Science* 51(6): 2597–2606.

Henderson, C. R. 1985. Best linear unbiased prediction of nonadditive genetic merits in noninbred populations. *Journal of Animal Science* 60(1): 111–117.

Heslot, N., Akdemir, D., Sorrells, M. E. and Jannink, J. L. 2014. Integrating environmental covariates and crop modeling into the genomic selection framework to predict genotype by environment interactions. *Theoretical and Applied Genetics* 127(2): 463–480.

Hickey, J. M., Dreisigacker, S., Crossa, J., Hearne, S., Babu, R. et al. 2014. Evaluation of genomic selection training population designs and genotyping strategies in plant breeding programs using simulation. *Crop Science* 54(4): 1476–1488.

Isidro, J., Jannink, J.-L., Akdemir, D., Poland, J., Heslot, N. et al. 2015. Training set optimization under population structure in genomic selection. *Theoretical and Applied Genetics* 128(1): 145–158.

Jan, H. U., Abbadi, A., Lücke, S., Nichols, R. A. and Snowdon, R. J. et al. 2016. Genomic prediction of testcross performance in canola (*Brassica napus*). *PLoS One* 11(1): e0147769.

Jarquín, D., Crossa, J., Lacaze, X., Du Cheyron, P., Daucourt, J. et al. 2014. A reaction norm model for genomic selection using high-dimensional genomic and environmental data. *Theoretical and Applied Genetics* 127(3): 595–607.

Jia, J. Y. and Jannink, J. L. 2012. Multiple-trait genomic selection methods increase genetic value prediction accuracy. *Genetics* 192(4): 1513–1522.

Jiang, Y. and Reif, J. C. 2015. Modeling epistasis in genomic selection. *Genetics* 201(2): 759–768.

Kaur, J., Akhatar, J., Goyal, A., Kaur, N., Kaur, S. et al. 2020. Genome wide association mapping and candidate gene analysis for pod shatter resistance in *Brassica juncea* and its progenitor species. *Molecular Biology Reports* 47: 2963–2974.

Khanzada, H., Wassan, G. M., He, H., Mason, A. S., Keerio, A. A. et al. 2020. Differentially evolved drought stress indices determine the genetic variation of *Brassica napus* at seedling traits by genome-wide association mapping. *Journal of Advanced Research* 24: 447–461.

Lado, B., Barrios, P. G., Quincke, M., Silva, P., Gutiérrez, L. et al. 2016. Modeling genotype x environment interaction for genomic selection with unbalanced data from a wheat breeding program. *Crop Science* 56(5): 2165–2179.

Lee, H., Chawla, H. S., Obermeier, C., Dreyer, F., Abbadi, A. et al. 2020. Chromosome-scale assembly of winter oilseed rape *Brassica napus*. *Frontiers in Plant Science* 11: 496.

Lin, Z., Hayes, B. J. and Daetwyler, H. D. 2014. Genomic selection in crops, trees and forages: A review. *Crop and Pasture Science* 65(11): 1177–1191.

Liu, S., Liu, Y., Yang, X., Tong, C., Edwards, D. et al. 2014. The *Brassica oleracea* genome reveals the asymmetrical evolution of polyploid genomes. *Nature Communications* 5(1): 1–11.

Liu, S., Fan, C., Li, J., Cai, G., Yang, Q. et al. 2016. A genome-wide association study reveals novel elite allelic variations in seed oil content of *Brassica napus*. *Theoretical and Applied Genetics* 129(6): 1203–1215.

Liu, X., Wang, H., Hu, X., Li, K., Liu, Z. et al. 2019. Improving genomic selection with quantitative trait loci and nonadditive effects revealed by empirical evidence in maize. *Frontiers in Plant Science* 10: 1129.

Liu, X., Wang, H., Wang, H., Guo, Z., Xu, X. et al. 2018. Factors affecting genomic selection revealed by empirical evidence in maize. *The Crop Journal* 6(4): 341–352.

Lopez-Cruz, M., Crossa, J., Bonnett, D., Dreisigacker, S., Poland, J. et al. 2015. Increased prediction accuracy in wheat breeding trials using a marker x environment interaction genomic selection model. *G3: Genes, Genomes, Genetics* 5(4): 569–582.

Lorenzana, R. E. and Bernardo, R. 2009. Accuracy of genotypic value predictions for marker-based selection in biparental plant populations. *Theoretical and Applied Genetics* 120(1): 151–161.

Lozada, D. N., Mason, R. E., Sarinelli, J. M. and Brown-Guedira, G. 2019. Accuracy of genomic selection for grain yield and agronomic traits in soft red winter wheat. *BMC Genetics* 20(1): 1–12.

Lyra, D. H., Granato, Í. S. C., Morais, P. P. P., Alves, F. C., dos Santos, A. R. M. et al. 2018. Controlling population structure in the genomic prediction of tropical maize hybrids. *Molecular Breeding* 38(10): 1–17.

Mageto, E. K., Crossa, J., Pérez-Rodríguez, P., Dhliwayo, T., Palacios-Rojas, N. et al. 2020. Genomic prediction with genotype by environment interaction analysis for kernel zinc concentration in tropical maize germplasm. *G3: Genes, Genomes, Genetics* 10(8): 2629–2639.

Mahmood, T., Ekuere, U., Yeh, F., Good, A. G., Stringam, G. R. et al. 2003. Molecular mapping of seed aliphatic glucosinolates in *Brassica juncea*. *Genome* 46(5): 753–760.

Manolio, T. A., Collins, F. S., Cox, N. J., Goldstein, D. B., Hindorff, L. A. et al. 2009. Finding the missing heritability of complex diseases. *Nature* 461(7265): 747–753.

Meuwissen, T. H., Hayes, B. J. and Goddard, M. E. 2001. Prediction of total genetic value using genome-wide dense marker maps. *Genetics* 157(4): 1819–1829.

Meuwissen, T. H. 2009. Accuracy of breeding values of 'unrelated' individuals predicted by dense SNP genotyping. *Genetics Selection Evolution* 41(1): 1–9.

Meuwissen, T., Hayes, B. and Goddard, M. 2016. Genomic selection: A paradigm shift in animal breeding. *Animal Frontiers* 6(1): 6–14.

Mrode, R., Ojango, J. M., Okeyo, A. M. and Mwacharo, J. M. 2019. Genomic selection and use of molecular tools in breeding programs for indigenous and crossbred cattle in developing countries: Current status and future prospects. *Frontiers in Genetics* 9: 694.

Nagaharu, U. and Nagaharu, N. 1935. Genome analysis in Brassica with special reference to the experimental formation of *B. napus* and peculiar mode of fertilization. *Japanese Journal of Botany* 7(7): 389–452.

Norman, A., Taylor, J., Edwards, J. and Kuchel, H. 2018. Optimising genomic selection in wheat: Effect of marker density, population size and population structure on prediction accuracy. *G3: Genes, Genomes, Genetics* 8(9): 2889–2899.

Panjabi-Massand, P., Yadava, S. K., Sharma, P., Kaur, A., Kumar, A. et al. 2010. Molecular mapping reveals two independent loci conferring resistance to *Albugo candida* in the east European germplasm of oilseed mustard *Brassica juncea*. *Theoretical and Applied Genetics* 121(1): 137–145.

Paritosh, K., Yadava, S. K., Gupta, V., Panjabi-Massand, P., Sodhi, Y. S. et al. 2013. RNA-seq based SNPs in some agronomically important oleiferous lines of *Brassica rapa* and their use for genome-wide linkage mapping and specific-region fine mapping. *BMC Genomics* 14(1): 1–13.

Paritosh, K., Yadava, S. K., Singh, P., Bhayana, L., Mukhopadhyay, A. et al. 2021. A chromosome-scale assembly of allotetraploid *Brassica juncea* (AABB) elucidates comparative architecture of the A and B genomes. *Plant Biotechnology Journal* 19(3): 602.

Qiu, D., Morgan, C., Shi, J., Long, Y., Liu, J. et al. 2006. A comparative linkage map of oilseed rape and its use for QTL analysis of seed oil and erucic acid content. *Theoretical and Applied Genetics* 114(1): 67–80.

Qu, C., Jia, L., Fu, F., Zhao, H., Lu, K. et al. 2017. Genome-wide association mapping and Identification of candidate genes for fatty acid composition in *Brassica napus* L. using SNP markers. *BMC Genomics* 18(1): 1–17.

Rahman, M., Mamidi, S., del Rio, L., Ross, A., Kadir, M. M. et al. 2016. Association mapping in *Brassica napus* (L.) accessions identifies a major QTL for blackleg disease resistance on chromosome A01. *Molecular Breeding* 36(7): 1–15.

Rakow, G. 2004. Species origin and economic importance of *Brassica*. *In*: Pua, E. C. and Douglas, C. J. [eds.]. *Brassica: Biotechnology in Agriculture and Forestry*. vol 54. Springer, Berlin, Heidelberg.

Ramchiary, N., Padmaja, K. L., Sharma, S., Gupta, V., Sodhi, Y. S. et al. 2007a. Mapping of yield influencing QTL in *Brassica juncea*: implications for breeding of a major oilseed crop of dryland areas. *Theoretical and Applied Genetics* 115(6): 807–817.

Ramchiary, N., Bisht, N. C., Gupta, V., Mukhopadhyay, A., Arumugam, N. et al. 2007b. QTL analysis reveals context-dependent loci for seed glucosinolate trait in the oilseed *Brassica juncea*: importance of recurrent selection backcross scheme for the identification of 'true' QTL. *Theoretical and Applied Genetics* 116(1): 77–85.

Riedelsheimer, C., Endelman, J. B., Stange, M., Sorrells, M. E., Jannink, J. L. et al. 2013. Genomic predictability of interconnected biparental maize populations. *Genetics* 194(2): 493–503.

Rio, S., Mary-Huard, T., Moreau, L. and Charcosset, A. 2018. Genomic selection efficiency and a priori estimation of accuracy in a structured dent maize panel. *Theoretical and Applied Genetics* 132(1): 81–96.

Roorkiwal, M., Jarquin, D., Singh, M. K., Gaur, P. M., Bharadwaj, C. et al. 2018. Genomic-enabled prediction models using multi-environment trials to estimate the effect of genotype x environment interaction on prediction accuracy in chickpea. *Scientific Reports* 8(1): 1–11.

Rousseau-Gueutin, M., Belser, C., Da Silva, C., Richard, G., Istace, B. et al. 2020. Long-read assembly of the *Brassica napus* reference genome Darmor-bzh. *GigaScience* 9(12): giaa137.

Rout, K., Yadav, B. G., Yadava, S. K., Mukhopadhyay, A., Gupta, V. et al. 2018. QTL landscape for oil content in *Brassica juncea*: analysis in multiple bi-parental populations in high and "0" erucic background. *Frontiers in Plant Science* 9: 1448.

Schopp, P., Müller, D., Wientjes, Y. C. and Melchinger, A. E. 2017. Genomic prediction within and across biparental families: Means and variances of prediction accuracy and usefulness of deterministic equations. *G3: Genes, Genomes, Genetics* 7(11): 3571–3586.

Sharma, R., Aggarwal, R. A., Kumar, R., Mohapatra, T., Sharma, R. P. et al. 2002. Construction of an RAPD linkage map and localization of QTLs for oleic acid level using recombinant inbreds in mustard (*Brassica juncea*). *Genome* 45(3): 467–472.

Shi, J., Zhan, J., Yang, Y., Ye, J., Huang, S. et al. 2015. Linkage and regional association analysis reveal two new tightly-linked major-QTLs for pod number and seed number per pod in rapeseed (*Brassica napus* L.). *Scientific Reports* 5(1): 1–18.

Snowdon, R. J. 2007. Cytogenetics and genome analysis in *Brassica* crops. *Chromosome Research* 15(1): 85–95.

Song, J. M., Guan, Z., Hu, J., Guo, C., Yang, Z. et al. 2020. Eight high-quality genomes reveal pan-genome architecture and ecotype differentiation of *Brassica napus*. *Nature Plants* 6(1): 34–45.

Spindel, J. E., Begum, H., Akdemir, D., Collard, B., Redoña, E. et al. 2016. Genome-wide prediction models that incorporate *de novo* GWAS are a powerful new tool for tropical rice improvement. *Heredity* 116(4): 395–408.

Sun, C., Wang, B., Yan, L., Hu, K., Liu, S., Zhou, Y. et al. 2016a. Genome-wide association study provides insight into the genetic control of plant height in rapeseed (*Brassica napus* L.). *Frontiers in Plant Science* 7: 1102.

Sun, C., Wang, B., Wang, X., Hu, K., Li, K. et al. 2016b. Genome-wide association study dissecting the genetic architecture underlying the branch angle trait in rapeseed (*Brassica napus* L.). *Scientific Reports* 6(1): 1–11.

Sun, F., Fan, G., Hu, Q., Zhou, Y., Guan, M. et al. 2017. The high-quality genome of *Brassica napus* cultivar 'ZS 11' reveals the introgression history in semi-winter morphotype. *The Plant Journal* 92(3): 452–468.

Sun, M., Hua, W., Liu, J., Huang, S., Wang, X. et al. 2012. Design of new genome-and gene-sourced primers and identification of QTL for seed oil content in a specially high-oil *Brassica napus* cultivar. *PLoS ONE* 7(10): e47037.

Tang, M., Zhang, Y., Liu, Y., Tong, C., Cheng, X. et al. 2019. Mapping loci controlling fatty acid profiles, oil and protein content by genome-wide association study in *Brassica napus*. *The Crop Journal* 7(2): 217–226.

[USDA] Foreign Agriculture Service. United States Department of Agriculture 2020.

VanRaden, P. M. 2008. Efficient methods to compute genomic predictions. *Journal of Dairy Science* 91(11): 4414–4423.

Wang, H., Wang, Q., Pak, H., Yan, T., Chen, M. et al. 2021. Genome-wide association study reveals a patatin-like lipase relating to the reduction of seed oil content in *Brassica napus*. *BMC Plant Biology* 21(1): 1–12.

Wang, T., Wei, L., Wang, J., Xie, L., Li, Y. Y. et al. 2020. Integrating GWAS, linkage mapping and gene expression analyses reveals the genetic control of growth period traits in rapeseed (*Brassica napus* L.). *Biotechnology for Biofuels* 13(1): 1–19.

Wang, D., Eskridge, K. M. and Crossa, J. 2011a. Identifying QTLs and epistasis in structured plant populations using adaptive mixed LASSO. Journal of Agricultural, Biological, and Environmental Statistics 16(2): 170–184.

Wang, X., Wang, H., Wang, J., Sun, R., Wu, J. et al. 2011b. The genome of the mesopolyploid crop species *Brassica rapa*. *Nature Genetics* 43(10): 1035–1039.

Wang, X., Yang, Z. and Xu, C. 2015. A comparison of genomic selection methods for breeding value prediction. *Science Bulletin* 60(10): 925–935.

Wang, X., Chen, L., Wang, A., Wang, H., Tian, J. et al. 2016. Quantitative trait loci analysis and genome-wide comparison for silique related traits in *Brassica napus*. *BMC Plant Biology* 16(1): 1–15.

Wang, X., Li, L., Yang, Z., Zheng, X., Yu, S. et al. 2017. Predicting rice hybrid performance using univariate and multivariate GBLUP models based on North Carolina mating design II. *Heredity* 118(3): 302–310.

Wang, X., Xu, Y., Hu, Z. and Xu, C. 2018. Genomic selection methods for crop improvement: Current status and prospects. *The Crop Journal* 6(4): 330–340.

Wassan, G. M., Khanzada, H., Zhou, Q., Mason, A. S., Keerio, A. A. et al. 2021. Identification of genetic variation for salt tolerance in *Brassica napus* using genome-wide association mapping. *Molecular Genetics and Genomis* 296(2): 391–408.

Werner, C. R., Qian, L., Voss-Fels, K. P., Abbadi, A., Leckband, G. et al. 2018a. Genome-wide regression models considering general and specific combining ability predict hybrid performance in oilseed rape with similar accuracy regardless of trait architecture. *Theoretical and Applied Genetics* 131(2): 299–317.112.

Werner, C. R., Voss-Fels, K. P., Miller, C. N., Qian, W., Hua, W. et al. 2018b. Effective genomic selection in a narrow-genepool crop with low-density markers: Asian rapeseed as an example. *The Plant Genome* 11(2): 170084.

Whittaker, J. C., Thompson, R. and Denham, M. C. 2000. Marker-assisted selection using ridge regression. *Genetics Research* 75(2): 249–252.

Wolc, A., Arango, J., Settar, P., Fulton, J. E., O'Sullivan, N. P. et al. 2016. Mixture models detect large effect QTL better than GBLUP and result in more accurate and persistent predictions. *Journal of Animal Science and Biotechnology* 7(1): 1–6.

Wu, J., Zhao, Q., Liu, S., Shahid, M., Lan, L. et al. 2016b. Genome-wide association study identifies new loci for resistance to Sclerotinia stem rot in *Brassica napus*. *Frontiers in Plant Science* 7: 1418.
Wu, Z., Wang, B., Chen, X., Wu, J., King, G. J. et al. 2016a. Evaluation of linkage disequilibrium pattern and association study on seed oil content in *Brassica napus* using ddRAD sequencing. *PLoS One* 11(1): e0146383.
Würschum, T. 2012. Mapping QTL for agronomic traits in breeding populations. *Theoretical and Applied Genetics* 125(2): 201–210.
Würschum, T., Abel, S. and Zhao, Y. 2014. Potential of genomic selection in rapeseed (*Brassica napus* L.) breeding. *Plant Breeding* 133(1): 45–51.
Xiao, Z., Zhang, C., Tang, F., Yang, B., Zhang, L. et al. 2019. Identification of candidate genes controlling oil content by combination of genome-wide association and transcriptome analysis in the oilseed crop *Brassica napus*. *Biotechnology for Biofuels* 12(1): 1–16.
Xu, L., Hu, K., Zhang, Z., Guan, C., Chen, S. et al. 2016. Genome-wide association study reveals the genetic architecture of flowering time in rapeseed (*Brassica napus* L.). *DNA Research* 23(1): 43–52.
Xu, Y., Li, P., Yang, Z. and Xu, C. 2017. Genetic mapping of quantitative trait loci in crops. *The Crop Journal* 5(2): 175–184.
Xuan, L., Yan, T., Lu, L., Zhao, X., Wu, D. et al. 2020. Genome-wide association study reveals new genes involved in leaf trichome formation in polyploid oilseed rape (*Brassica napus* L.). *Plant, Cell & Environment* 43(3): 675–691.
Yabe, S., Hara, T., Ueno, M., Enoki, H., Kimura, T. et al. 2018. Potential of genomic selection in mass selection breeding of an allogamous crop: an empirical study to increase yield of common buckwheat. *Frontiers in Plant Science* 9: 276.
Yadava, S. K., Arumugam, N., Mukhopadhyay, A., Sodhi, Y. S., Gupta, V. et al. 2012. QTL mapping of yield-associated traits in *Brassica juncea*: meta-analysis and epistatic interactions using two different crosses between east European and Indian gene pool lines. *Theoretical and Applied Genetics* 125(7): 1553–1564.
Yan, W., Zhao, H., Yu, K., Wang, T., Khattak, A. N. et al. 2020. Development of a multiparent advanced generation intercross (MAGIC) population for genetic exploitation of complex traits in *Brassica juncea*: Glucosinolate content as an example. *Plant Breeding* 139(4): 779–789.
Yang, J., Liu, D., Wang, X., Ji, C., Cheng, F. et al. 2016. The genome sequence of allopolyploid *Brassica juncea* and analysis of differential homoeolog gene expression influencing selection. *Nature Genetics* 48(10): 1225–1232.
Yang, Q., Fan, C., Guo, Z., Qin, J., Wu, J. et al. 2012. Identification of FAD2 and FAD3 genes in *Brassica napus* genome and development of allele-specific markers for high oleic and low linolenic acid contents. *Theoretical and Applied Genetics* 125(4): 715–729.
Yang, Y., Shen, Y., Li, S., Ge, X. and Li, Z. et al. 2017. High density linkage map construction and QTL detection for three silique-related traits in *Orychophragmus violaceus* derived *Brassica napus* population. *Frontiers in Plant Science* 8: 1512.
Zhang, A., Wang, H., Beyene, Y., Semagn, K., Liu, Y. et al. 2017. Effect of trait heritability, training population size and marker density on genomic prediction accuracy estimation in 22 bi-parental tropical maize populations. *Frontiers in Plant Science* 8: 1916.
Zhang, Y., Thomas, C. L., Xiang, J., Long, Y., Wang, X. et al. 2016. QTL meta-analysis of root traits in *Brassica napus* under contrasting phosphorus supply in two growth systems. *Scientific Reports* 6(1): 1–12.
Zhao, Y., Gowda, M., Liu, W., Würschum, T., Maurer, H. P. et al. 2012. Accuracy of genomic selection in European maize elite breeding populations. *Theoretical and Applied Genetics* 124(4): 769–776.

Zhao, Y., Zeng, J., Fernando, R. and Reif, J. C. 2013. Genomic prediction of hybrid wheat performance. *Crop Science* 53(3): 802–810.

Zhao, Y., Mette, M. F., Gowda, M., Longin, C. F. H., Reif, J. C. et al. 2014. Bridging the gap between marker-assisted and genomic selection of heading time and plant height in hybrid wheat. *Heredity* 112(6): 638–645.

Zou, J., Zhao, Y., Liu, P., Shi, L., Wang, X. et al. 2016. Seed quality traits can be predicted with high accuracy in *Brassica napus* using genomic data. *PLoS One* 11(11): e0166624.

8

Current Status of Genomic Selection in Oilseed Crops

Megha Sharma,[1] *Pooja Kaushik,*[1] *Ani A Elias*[2] and *Shailendra Goel*[1,*]

ABSTRACT

Major oilseeds such as soybean, oil palm, rapeseed, and sunflower contribute significantly to human as well as animal nutritional needs. The increase in oilseed production is, however, only marginal due to limited cultivation areas and climate changes. Genomic selection (GS) has led to significant genetic gain and acts as a game changer by speeding up the breeding programs mainly through shortening the breeding cycle and increasing the selection intensity. With a continuous decrease in the cost of genomic data generation and the development of statistical tools and software, GS became more accessible to breeders. In this chapter, we examine the status of GS in major as well as minor oilseed crops. Most of the studies on oilseed crops emphasize factors affecting the GS of oil production, complex traits, training population size, models, genomic relationships, marker type and density, and heritability. Also, the ability of GS to predict the parental lines indicates its potential in hybrid breeding as demonstrated in sunflower. We infer that comprehensive breeding programs are required for minor oilseed crops also. This way, GS will help in meeting global non-nutritional demands such as medicine development as well.

[1] Department of Botany, University of Delhi, New Delhi, India - 110007.
[2] Institute of Forest Genetics and Tree Breeding, Coimbatore, Tamil Nadu, India – 641002.
* Corresponding author: Shailendragoel@gmail.com

Introduction

Oilseed crops are crucial food elements for the growing global population. With the human population anticipated to exceed 9 billion by 2050, oilseed crops are expected to provide 44 percent of the required calories (www.fao.org). Agriculturally, a significant area (more than 180 million hectares), is occupied by oilseeds such as sunflower, soybean, and rapeseed (Silaeva, 2016). There are three types of oilseed crops: annual or biennial crops such as soybean, sunflower, and rapeseed, and perennial tree crops such as oil palm. Embryos are a by-product of the third category, which includes cotton and corn germ. The rest of the oilseeds are minor players in global commerce, although they are important in local markets; among these essential crops are safflower and flax. About 40 different oilseeds provide oil fit for human consumption, but only a few are significant in world trade and supply. Besides consumption as human food, oilseeds also serve as an essential source of dietary proteins for animals in unprocessed or processed forms such as cakes or meals. For these reasons, oilseed crops have been grown all over the world and are considered economically important. However, considering the increasing demand, the oilseed production in the world has increased marginally. In the world market, the countries of the European Union (EU) rose as the main producers of oilseeds, accounting for more than 50% of the oil yield (Abakumov, 2012). Soybean stands as a crucial oilseed crop with the USA, Brazil, and Argentina as the primary producers. Likewise, with Canada and China being the largest producers, rapeseed is second in the world oilseed production. The third spot is occupied by sunflower in the rankings for the global production of oilseeds. The EU countries, the Russian Federation and Ukraine are the top producers of sunflower, accounting for more than 70% of its production (Baigot and Glotova, 2017). To fill the demand-supply gap, genetic improvement of the oilseed crops plays a significant role.

The primary goal of plant breeding is the application of genetic principles to improve plants that are economically or aesthetically desirable to humans. Enhancement of seed and oil production, quality of oil according to its application, i.e., edible or industrial purposes, breeding of varieties that fit in different cropping systems, and breeding biotic and abiotic stress-resistant/tolerant types are the major goals in oil crop development (Yadava et al., 2012). Pure-line selection and mutation breeding are two traditional plant breeding strategies that have resulted in the production of oilseed crop variants with more seed oil. The genetic diversity available in a crop species, however, limits the potential for enhancing oil characteristics using these approaches (Bernardo and Yu, 2007). Furthermore, the oil yield is a quantitative trait regulated by the environment and controlled by multiple loci (Rahman et al., 2013). Although marker-assisted selection (MAS) has played a significant role in plant breeding for disease and pest resistance, it is difficult to apply it to quantitative traits such as grain yield and seed oil content (Azevedo Peixoto et al., 2017). Further, the environment has a large effect on these traits, resulting in small to moderate heritability. Thus, genomic selection (GS) is a promising method that utilizes genome-wide markers and can be applied to oilseed crops widely in the selection of complex traits in plant breeding programs.

GS helps in the selection of agronomically superior quantitative traits. In plant breeding programs, it promotes genetic gain or efficiency (Bernardo and Yu, 2007; Gaynor et al., 2017; Hickey et al., 2014). It does this by shortening the breeding cycle, improving selection precision, and boosting selection intensity, all of which are important elements in the breeder's equation. The training of a GS model requires a phenotyped and genotyped training population (TP). The genomic estimated breeding values (GEBVs) of individuals in a breeding population that has been genotyped but not phenotyped are then predicted using this approach. After that, the GEBVs are used to select superior lines. The correlation between predicted values through models and the observed phenotypes provides prediction accuracy. The GS accuracy is influenced by multiple factors such as genetic architecture, heritability and sample size that will be discussed in detail.

The review of the literature indicates the increased application of GS in oilseed crop with every passing year. There were only 3 research papers in 2001 when GS was first introduced, which has increased to around 563 in 2020 (Fig. 1). Examining the current status of GS in the oilseed crops shows that more emphasis had been given to major oilseeds crops and minor seeds crops were marginalized due to lack of a comprehensive breeding program and limited availability of genomic resources. The minor crops have the potential to bridge the supply-demand gap of edible oil and the application of GS can fast track the improvement and revolutionize the nutritional security in traditional agricultural belts.

Factors affecting GS in oilseed crops

The correlation between the estimated breeding values and true breeding values is said to be the accuracy of estimated breeding values. In practice, observable phenotypes are frequently used to calculate accuracy as a proxy for true breeding values when GEBVs are evaluated (Lin et al., 2014). The accuracy is important since it is directly proportional to the genetic gain. There are many factors which affect the accuracy of the GS and are interrelated in a complex and comprehensive manner (Zhang et al., 2016).

One of the major factors which determine the accuracy is the selection of the statistical methods. Relationship based methods and marker-effect based methods are two types of statistical models. In the relationship-based

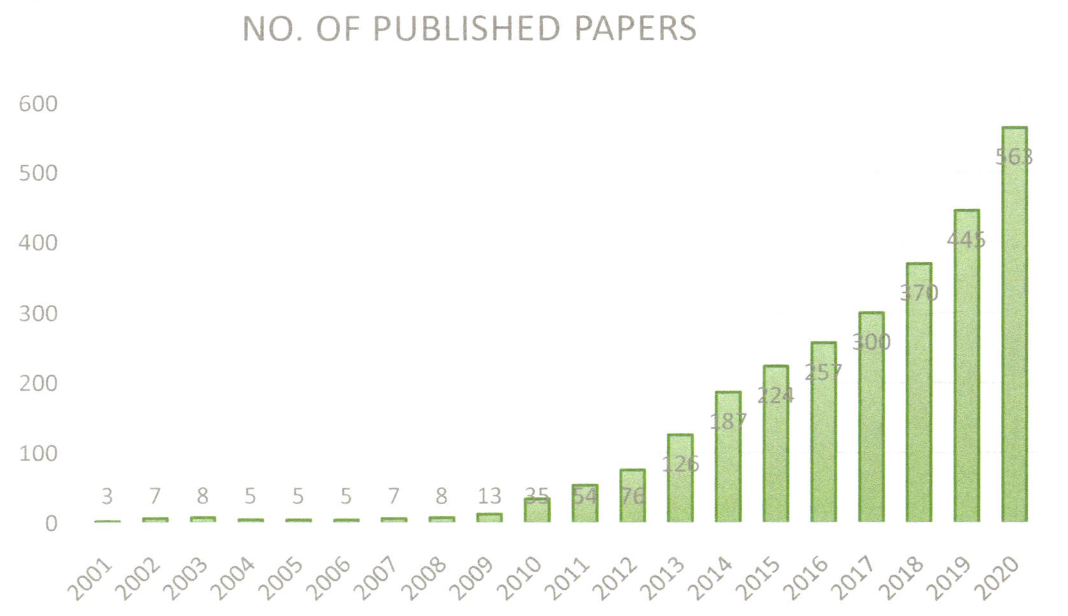

Fig. 1. The figure depicts the year-wise distribution of the number of publications from the year 2001 to 2020. The data was retrieved from google scholar by using the keywords "genomic selection in oilseed crops" using different permutations and combinations.

method, breeding values are predicted by using relationship information rather than calculating the impacts of genetic markers. The relationship matrix is derived from pedigree information using the traditional best linear unbiased prediction (T-BLUP) technique. It was often used in the breeding of cattle (Hidalgo et al., 2015). However, the modes of reproduction limit the usage of pedigree information based on BLUP in plant breeding (Hickey et al., 2014). In comparison, the Genomic Best linear unbiased prediction (GBLUP) method uses a genomic information-based relationship matrix. It was widely used in both animal and plant breeding because it provided more accurate relationship coefficients among individuals and enhanced the estimation accuracy of breeding values (VanRaden, 2008). The marker effect-based approaches, on the other hand, assess the effects of genetic markers first, then add them together to get each individual's estimated breeding values (EBV). The marker effect-based approaches include the Ridge regression BLUP (RR-BLUP) method and Bayesian alphabet methods (Bayes A, Bayes B, Bayes C). In RR-BLUP, it is assumed that all markers have the same genetic variance and have been proved to be equivalent with the GBLUP method (Goddard, 2009; Hayes et al., 2009). Bayes A method assumes that the variance of proportion of all marker effects follows an inverse-chi-square distribution (Meuwissen et al., 2001). In BayesB, a proportion (π) of markers contribute to the target trait ($\pi > 0$), and the variance of marker effects follows an inverse-chi-square distribution (Meuwissen, 2009). BayesC, BayesCπ, BayesLASSO, make a better optimization of π and assigned distinct distributions for the variances of marker effects (Habier et al., 2011; Yi and Xu, 2008). A large number of methods can be chosen to make predictions in GS: penalized regression methods, Bayesian methods, and reproducing kernel Hilbert space methods are the most popular tools. Various modeling approaches described above have been used in the GS of the oilseed crops. For instance, in oil palm various studies exploit the GBLUP models (Cros et al., 2015; Ithnin et al., 2017; Kwong et al., 2017; Wong and Bernardo, 2008) and were used in each parental group separately, with data records consisting of parental performances in crosses with the other group, i.e., general combining ability (GCA) or testcross phenotypic means and parent genotypes (Nyouma et al., 2019).

The choice of the model affects predictive performance, but the comparison of various models for genomic prediction (GP) has established the critical role of **genetic architecture** (de Los Campos et al., 2010; Howard et al., 2014). Various simulation studies suggest that the model chosen for prediction depends on genetic architecture, for example, a simulation study (Daetwyler et al., 2010) showed that the accuracy of G-BLUP is not affected by the number of QTLs; whereas the predictive performance of BayesC was greatly affected by the genetic architecture and its accuracy was higher when the number of QTLs was low and decreased with an increased number of QTL.

Using RR-BLUP, Würschum et al., 2013 observed that traits which possess less complex genetic architecture show higher prediction accuracies in the *Brassica napus*. The prediction accuracies were lowest for the glucosinolate content and grain yield which are complex quantitative traits and highest for plant height on the other hand.

Another crucial factor is the **heritability of the trait**. Ideally, high-heritability traits are positively correlated with higher GEBV prediction. Zhao et al. (2013) noted the accuracy for yield (low heritability) and grain moisture (high heritability) in maize were 0.58 and 0.90, respectively. However, it is not always true as many agronomically important traits have low heritability but show high prediction accuracy (Heffner et al., 2011).

Furthermore, even though many influencing factors are critical, increasing **population size** generally improves the accuracy of GEBVs. For instance, decreasing the training population size from 288 to 198 and 96 reduced the average GS prediction accuracy by 11 and 30 percent and conventional marker-assisted selection prediction accuracy by 24 and 35 percent, respectively (Heffner et al., 2011).

In addition to these factors, the availability of **high-density markers** and cost-effective genotyping is important to provide faster genetic gain than conventional selection methods. Solberg et al. (2008), found that two to three times higher density of SNP markers was required compared to microsatellite markers to attain similar accuracy. However, empirical and simulation studies in crops such as maize and soybean revealed that increasing the marker density beyond a threshold level does not significantly improve the accuracy and genetic gain (Ramasubramanian and Beavis, 2020).

The accuracy of GS is influenced by the **relatedness** between the phenotyped individuals in the training set and the individuals that are to be predicted (Clark et al., 2012; 2012; Habier et al., 2007; Hickey et al., 2014; Liu et al., 2016; Meuwissen, 2009; Zhang et al., 2017). Small number of close relatives and very large numbers of distant relatives enable accurate predictions (Edwards et al., 2019). Small or modest numbers of distant relatives do not enable accurate predictions, as they share only a small proportion of the genome with the selected candidates and thus provide less reliable predictions (de Los Campos et al., 2013). Finally, the training set should also comprise a diverse set of individuals to produce reliable predictions (Pszczola and Calus, 2016), as supported by research studies in both cattle (Jenko et al., 2017) and barley (Neyhart et al., 2017).

In conclusion, the accuracy of GEBVs is affected by several factors. Oilseed crops are complex crops where most of the oil-related traits are quantitative in nature and effects of factors on GS were studied in different oilseed crops. Therefore, species-specific strategies need to be developed for

GS taking into account reproduction system, generation time, genome structure, harvested organs, breeding purposes, and environmental interactions.

Genomic selection in major oilseed crops

Soybean

Soybean (*Glycine max*) is a leading crop in the world's oilseed cultivation scenario owing to its high productivity and profitability (Simanjuntak et al., 2020). The current global oil production of soybean is 352 million tons (https://www.fao.org/faostat/en/#data/QC) which is expected to increase to 452 million tons in the next decade (OECD, 2020). It is one of the most traded agricultural products worldwide because of the varied types of consumption, ranging from food, animal feed, medicine, and other industrial use (Freitas et al., 2011).

Due to the high nutrition content of soybean seed (42.1% protein & 19.2% oil content) as well as its multitude of uses, constant efforts have been made for its improvement through conventional and modern breeding techniques (Duhnen et al., 2017). Most breeding efforts in soybean mainly focused on eliminating major problems of soybean oil such as low oxidative stability, off-flavor, and rancidity which required modification of fatty acids such as the oleic acid composition of soybean oil to improve its quality (O'Shea et al., 2020). However, the conventional breeding approach suffers from linkage drag and offers limited scope in the development of improved soybean genotypes in a shorter time. These limitations directed the soybean improvement program towards exploring the prospects of GS.

Over the last two decades a large number of SSR and SNP markers had been generated in soybean. To study the seed composition trait, SSR markers have been utilized extensively (Jarquín et al., 2014; Wang et al., 2014; Warrington et al., 2015). Also, SSR markers were used for the detection of QTLs for seed oil, protein, and seed size (Hyten et al., 2004) and fine mapping of soybean protein QTL on chromosome 20 (Nichols et al., 2006). A database for SSR markers which consists of 33,000 markers was developed through information from whole-genome sequencing (Song et al., 2010). The presence of well-annotated genome sequences has enabled the detection of the SNP (Singh et al., 2013) for soybean improvement. Different studies including QTL, GWAS, and whole genome resequencing (WGRS) had already summarized the chromosomal location of the trait. Most of the QTLs identified for seed composition are located on chromosomes 20, 15, 6, and 5. In soybean, meta-QTLs analyses have been performed for seed composition traits. For example, meta-QTLs analysis was performed on seed oil content (Reif et al., 2013) which gives the basis for gene mining and also refines soybean genetic maps. The major challenge of the phenotyping of the soybean is that its fatty acid composition is affected by environmental factors (Bellaloui et al., 2015) and could not be resolved by the conventional MAS approach. Hence, GS has the potential to address this challenge in near future.

For soybean, various studies had reported GS accuracy for production traits, development traits, or resistance to pathogens (Bao et al., 2014; Jarquín et al., 2014; Zhang et al., 2016). The first study for examining the potential of GS in soybean was done by Jarquin et al. (2014). They performed GBS on 301 elite soybean lines and scored 16, 502 SNPs with a minor allele frequency (MAF) > 0.05 and a percentage of missing values ≤ 5%. The prediction accuracy for grain yield was estimated to be 0.64, indicating a good potential of GS for the trait in soybean. Likewise, primary embryogenesis capacity in soybean was predicted using 80 SSR markers on the 126 recombinant inbred lines (Shu et al., 2013a). In that report, a high correlation (r^2 = 0.78) has been observed among the GEBVs and the phenotypic value. In another study, 79 SCAR markers were used on the 288 cultivars. The coefficient of correlation between GEBV and the phenotypic value was 0.90, which is higher than that in the previous report (r^2 = 0.79) (Crossa et al., 2010). Later (Shu et al., 2013b), applied a linear unbiased predictor and Bayesian linear regression models on the SCAR markers were used for the identification of the markers associated with the hundred-seed weight. Both GS models showed a good prediction performance with an accuracy value as high as 0.904.

Bao et al. (2014) genotyped 282 genotypes of soybean using a genome-wide panel of 1536 SNPs for investigating association of resistance to soybean cyst nematode (SCN). GS using the full marker set produced average prediction accuracy ranging from 0.59 to 0.67, which was significantly more accurate than MAS strategies using two rhg1-associated DNA makers. The number of markers and the accuracy of the GS were found to be related in this study. As the number of markers increased, the GS accuracy increased. For example, the effect of the training population size on accuracy began to plateau around 100 individuals, but accuracy steadily increased until the largest possible size of 251 was used in this analysis. Their results also suggest that standard additive G-BLUP models can be used on unfiltered, imputed GBS data without accuracy loss.

Thus, the GS in the soybean holds a good potential to expedite genetic gain for yield in soybean breeding programs (Cloutier et al., 2009; Duhnen et al., 2017). Through the few GS studies, it could be seen that implementation of the GS in the soybean is rewarding.

Sunflower

Sunflower (***Helianthus annuus***) is the second most important crop based on hybrid breeding, after maize (Seiler et al., 2017). It is used for its seed oil and seeds are also used in confectioneries as well as snacks. Sunflower is at number four with 12% of the global production of vegetable oil, after palm oil, soybean, and canola oil (Rauf et al., 2017). Earlier in 1970, the sunflower production was based on open-pollinated seeds, although it switched to hybrid breeding with the discovery of the CMS (cytoplasmic male sterility) lines. The development of sunflower hybrids converted it into a major viable oilseed crop worldwide and encouraged the foundation of numerous public and private breeding centers (Seiler et al., 2017). In the recent year, there has been worldwide research in the identification of markers for marker-assisted selection and also the establishment of high throughput technologies.

Over the years, various genetic resources had been generated and used for MAS in sunflowers. The improvement of the sunflower can be enhanced when available plant genetic resources are combined with the modern molecular tools for GWAS and GS (Dimitrijevic and Horn, 2018). However, the availability of genomic resources is very recent in sunflower as compared to other crops (Fig. 2).

GS seems promising in hybrid breeding of self-pollinating crops (Zhao et al., 2015), especially if little is known about the heterotic pool. It can predict performance through a model that is not necessarily trained for all hybrid parents. And thus it can solve the issue of an unknown parent in GCA (Dimitrijevic and Horn, 2018; Mangin et al., 2017). In sunflower, the prediction models so far have been based on RFLP marker, AFLP marker, and GBS (Baute et al., 2016; Celik et al., 2016; Ma et al., 2017; Qi et al., 2011, p. 017).

Reif et al. (2013) predicted the hybrid performance of the 104 intergroup and 133 intragroup hybrids with their parental lines. The prediction was based on the 572 AFLP markers on the parental lines. Different traits including oil yield, and oil content were tested in four different environment locations. Their results indicated that even if no information were present on the GCA ability of the parental lines, the GS was accurate in the case of the closely related parents while that for the distant parents was more difficult. Mangin et al. (2017a) used sequencing data for the genomic prediction of hybrid oil content for an incomplete factorial design consisting of 36 CMS lines and 36 restorer lines and compared the prediction accuracy of GS and classical GCA models in sunflower. Multi-environmental field trials were performed to characterize 452 sunflower hybrids of the panel about hybrid performance in oil content, which represents a primarily additive trait with high heritability. In addition, all 72 parental lines were sequenced to obtain genome-wide SNP markers (Mangin et al., 2017a). Genomic predictions were then made for missing hybrids and hybrid combinations lacking information for one parental line. Therefore, GS led to a considerable improvement in breeding efficiency even when little was known about the GCA of one or both parental lines (Mangin

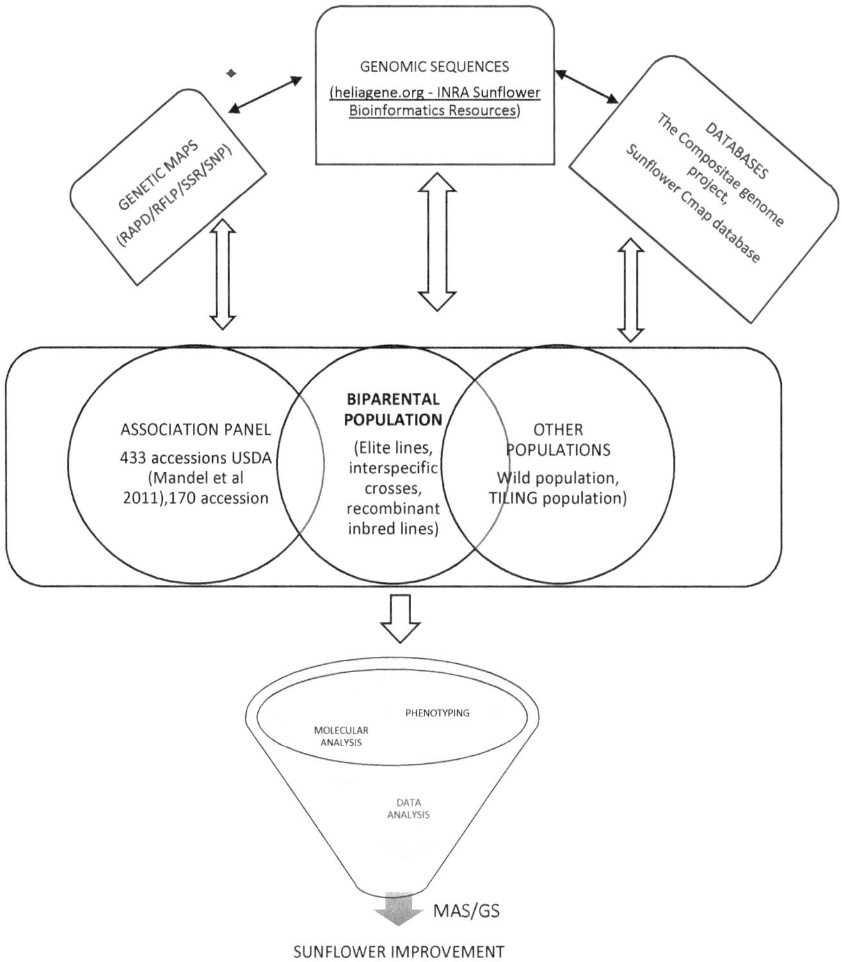

Fig. 2. Schematic general overview of the resources available in MAS and GS for sunflower breeding program (adopted from Dimitrijević and Horn, 2018).

et al., 2017a). These initial steps in GS concerning hybrid performance and hybrid oil content emphasized the potential of the method in speeding-up sunflower breeding programs in the future (Dimitrijevic and Horn, 2018).

Oil palm

Oil palm (*Elaeis guineensis*) is a diploid, monoecious, and allogamous perennial crop with high GS potential (Cros, 2015). Oil palm accounted for the annual production of more than 65 million tons of palm oil (GAIN Report, USDA Foreign Agriculture Service 2018: http://gain.fas.usda.gov/). It is one of the most productive oil crops in the world. With 35% of the world's vegetable oil production in 2015, the steady increase is an upward trend seen since the nineties (USDA, 2018).

Genomic improvement in oil palm started in the 1920s through mass selection followed by approaches such as modified reciprocal recurrent selection type (MRRS) which led to the hybrid cultivars. As MAS can shorten the breeding cycle as well as increase the selection intensity, it is widely applied in oil palm (Billotte et al., 2010; Ting et al., 2018; Tisné et al., 2015). However, MAS has limitations during analyses of complex traits such as yield which are controlled by multiple genes with small effects, especially in the case of small population sizes leading to overestimation of the strong QTLs and failure to exploit weak QTLs (Muranty et al., 2014). Estimation of genetic values with advanced statistical tools and the use of genomic data in the breeding pipeline are the two new changes that the current period of genetic improvement is witnessing (Nyouma et al., 2019). Thus, the most efficient genomic approach for quantitative traits after considering the above two resources is GS (Meuwissen et al., 2001). Its practical implementation was made possible by progress in genomics, in particular in next generation sequencing (NGS) and high throughput genotyping (Varshney et al., 2017).

Oil palm is turning into a model plant for GS studies among the perennial crops with the highest number of published studies among oilseed crops. In oil palm, GS had already been used for: (i) parent selection of hybrid crosses for yield traits, (ii) evaluation of its ability to reduce the length of breeding cycles by avoiding field trials in some cycles, (iii) increase in selection intensity by its application to a large number of candidates (Nyouma et al., 2019).

Although GS generally uses single nucleotide polymorphism markers (SNPs) in training the model, throughout time various molecular markers have been generated in oil palm. From the 1980s, Restricted Fragment Length Polymorphism (Botstein et al., 1980), Random Amplified Polymorphic DNA (Williams et al., 1990), Amplified Fragment Length Polymorphisms (Vos et al., 1995), short sequence repeats (SSRs) (Litt and Luty, 1989; Edwards et al., 1991), SSRs palm (López et al., 2004; Myint et al., 2019; Ting et al., 2014) were generated. Recently, a database for oil palm's microsatellite has been established (Babu et al., 2017). To analyze the genetic diversity, population structure, linkage mapping, and QTL mapping for important traits, SSR markers have been widely used (Wang et al., 1998). Apart from SSR, SNP can be used in oil palm which can be genotyped. The advancement of NGS technologies, and the substantial reduction in sequencing costs, made SNP more attractive for researchers (Davey et al., 2011; Bai et al., 2017). SNP can be detected using over 40 different methods (Kwong et al., 2016), which also includes genotyping with microarray, genome by sequencing (GBS). For oil palm, over a million SNPs have been generated recently (Jin et al., 2016; Kwong et al., 2016).

For the yield component, the effect of the number of markers has been studied through three different studies in oil palm. During the prediction of the performance of unevaluated hybrids, GS accuracy was found to be plateauing between 500 and 2000 SNPs markers (Cros et al., 2017) and between 200 and 400 SNP markers (Kwong et al., 2017). The SNPs required were lower in the latter case due to the selection of SNPs based on the GWAS study. However, during the prediction of GCA for progeny-tested individuals, Marchal et al. (2016) demonstrated that GS accuracy plateaued with 160 SSRs in group A palm (Deli/Asian palm known to have many branches) and 90 SSRs in group B (African oil palm known to have a few branches). Therefore, the marker density requirement for maximizing GS accuracy depends on the marker type, sampling method, trait, and population structure. Due to the higher inbreeding rate in oil palm, the marker density requirement in this crop is relatively low. The simulation studies have shown that two cycles of selection increased the 'per cycle response' by more than 10%, mainly as a result of higher selection accuracy (Cros et al., 2017).

Thus, GS is the new-age disruptive approach toward genetic improvement of the oil palm. However, further studies are required to make it more efficient for all yield components, which should include prediction model improvement as well as training population optimization. Also, the use of the multi-omics data in the GS modeling should be considered. However, along with the genetic improvement of yield components, other factors such as pest and prey resistance can make a huge difference in oil palm production in the coming years.

Genomic selection in minor seed crops

According to the FAO (2019), the total amount of oil production is around 520.7 million tons of which 80% of the production is constituted by soybean, canola, oil palm, and sunflower. Besides major crops, some such as linseed, safflower, groundnut, and cottonseed are the minor crops that are exploited for their oilseeds. These crops have unique properties and play a major role in traditional uses and the local economy. These minor oilseeds represent untapped genetic resources and do have the potential to bridge the gap between the supply and demand of edible oil. However, GS could not be exploited in many of the minor oilseed crops in all these years, and thus due to lack of literature, this part of the chapter focuses on flax (linseed) and safflower.

Flax

Flax (*Linum usitatissimum*) is a self-pollinating crop. It is an important multi-purpose crop, majorly used for its oil and fiber. The long-term domestication of flax resulted in two major morphotypes, linseed and fiber. The linseed morphotype is rich in oil (40–50%) containing five main fatty acids: palmitic

(PAL, C16:0, ~ 6%), stearic (STE, C18:0, ~ 2.5%), oleic (OLE, C18:1Δ9, ~ 19%), linoleic (LIO, C18:2Δ9, 12, ~ 13%), and linolenic (LIN, C18:3Δ9, 12, 15, ~ 55%) (Diederichsen et al., 2013). Linseed is rich in omega-3 fatty acids due to its higher linseed content. It is useful in mitigating heart disease by reducing the blood cholesterol level in humans (Green, 1986). These attributes also make it ideal as an industrial oil for use in paints, linoleum flooring, inks, soaps, and varnishes. From 2.16 million tonnes in 2010 to 3.4 million tons of oil in 2019, linseed production in the global market shows a steady increase over this decade (FAOSTAT, 2019). Breeding effort can potentially spike the oil production in the near future.

The major objectives of breeding in linseed include the development of varieties with high yield, high oil content, modified linseed content, increased straw yield, the fiber content in straw, fiber quality, and resistance to biotic and abiotic stresses. Most of these traits are quantitative thus breeding approaches such as MAS and GWAS were exploited for the dissection of these traits. However, GS can act as a better alternative in flax breeding.

In the last decade, flax genomics show significant progress such as the generation of a large number of molecular markers including SSR and SNP as well as quantitative trait loci (QTL) linked with important agronomic traits (You et al., 2019). More than 1,400 SSR markers have been developed using EST libraries (Cloutier et al., 2009). SSR markers have been used in flax for the construction of its genetic maps (Soto-Cerda et al., 2012; Cloutier et al., 2009), assessing genetic diversity (Soto-Cerda et al., 2012) QTL mapping (Cloutier et al., 2009), and association studies (Soto-Cerda et al., 2014). A consensus map is generated from three bi-parental populations by the incorporation of 770 markers. The map had a total length of 1551 cM with a mean marker density of one marker every 2 cM and covered an estimated 74% of the predicted flax genome size of 370 Mb (Cloutier, 2012). A large region of the flax genome is covered by these markers. Cloutier used the same set of markers for identifying the QTLs associated with the oil-producing gene. Soto-cerda et al. (2012) used them in association mapping approaches and to form a flax core collection that is used for identifying the QTL for seed quality traits. Also, You (2019) performed a genome-wide selective sweep scan for selection signatures and detected 114 genomic regions that accounted for 7.82% of the flax pseudomolecule. These overlapped with 11 GWAS-detected genomic regions associated with 18 QTL for 11 traits. You et al. (2016) evaluated GS in three bi-parental populations using this genomic information for seed yield and quality. In this study, the three factors, i.e., GS model, markers, and population structure were evaluated. Out of the three GS models: RR-BLUP, Bayesian LASSO (BL), and Bayesian ridge regression (BRR), BRR outperformed the other two. Out of the two types of markers used: full (set differed across the populations) and common (common in three populations) more variations were exhibited by the full SSR markers in the best-fitted models.

GS based on genome-wide QTL have the potential to predict superior individuals when it comes to complex traits. Therefore, GS along with the breeding approaches such as double hybrid would facilitate pyramiding of superior alleles of targeted traits and the development of the superior cultivar lines possible.

Safflower

Safflower (*Carthamus tinctorius* L.) is a member of the family Asteraceae. It is self-pollinated with a haploid genome size of 1.4 GB with 2n = 24 chromosomes (Kumari et al., 2017). It has nutritionally desirable seed oil with a high level of nutritionally desirable fatty acids, which is unique among oilseed crops. It is used as a coloring dye, food flavorings agent, and as a source of plant-derived pharmaceuticals (Bowers et al., 2016). Compared to other oil crops, safflower is underutilized due to its low oil content, low seed yields, susceptibility to different diseases, and pests which decrease the quality and productivity (Ali et al., 2019). Currently, safflower is cultivated in 23 countries producing 0.62 million tons of seeds of safflower (FAOSTAT). Despite its shortcomings, safflower is cultivated due to its adaptability to drought and saline conditions (Weiss, 1983), superior quality oil, and its multi-purpose usage.

Genetic improvement through breeding in safflower is commonly aimed at improving yield, oil content, and other agronomic characteristics, including resistance to diseases, insects, and abiotic stresses. The indirect selection which includes the number of capitula per plant, capitulum diameter, and grains per capitulum (Arslan, 2007) has been used for improving grain yield, grain oil content, and oil yield (Golparvar, 2011). However, all these traits are quantitative and complex, and thus the most popular breeding approach MAS seems inefficient. Thus, approaches such as GS which estimate all markers according to their contribution can be useful and efficient.

Being a domesticated crop, safflower has gone through a genetic bottleneck reducing its genetic diversity. However, sufficient genetic diversity is available due to different diversity centers. The safflower genetic diversity has been estimated using several molecular markers such as RAPD (Amini et al., 2008; Khan et al., 2009), Inter-Simple Sequence Repeat or ISSR (Yang et al., 2007) AFLP (Kumar et al., 2015), SSRs (Ambreen et al., 2015; Lee et al., 2014), iPBS-retrotransposon (Ali et al., 2019), and SNP. These studies provide evidence of genetic diversity in the global germplasms of safflower. Also, the generation of the chromosomal scale reference genome of the safflower (Wu et al., 2021) allows the generation of more genetic resources in the safflower which could be used for training the GS model shortly.

The first study of GS in safflower was reported in 2020 (Zhao et al., 2021). Zhao et al. (2021) used single and multi-site models on the accessions from the Australian grain gene bank to study their population structure and assessed the potential of genomic prediction (GP) to evaluate yield and grain-related traits. In their studies of a single-site model, for all traits, the prediction accuracies of GBLUP ranged from 0.21 to 0.86. The prediction accuracy is consistent with estimated genomic heritability (h^2), which varied from low to moderate across traits. A low level of G × E interaction was observed. The GS is a new arena for safflower; with the development of more markers, the introgression of the desired safflower traits into the breeding lines from the gene bank germplasms will be facilitated by this tool.

Conclusion

Genomic selection has been widely used in the improvement of oil crops. GS has clear advantages over QTL-MAS as well as GWAS studies in the case of oil production as it is a quantitative trait. In oil crops, the major focus of improvement through GS is on the major crops such as soybean, sunflower, and oil palm. However, minor oil crops such as safflower and linseed which have a great potential in the oil industry are slowly gaining pace through GS. The exploitation of the underutilized crops through GS can lead to unveiling their huge potential as oil crops. Thus, the crop improvement in oil crops with the modern breeding techniques and tools can increase healthy oil production for multi-purpose use leading to higher nutritional standards of end-users.

References

Abakumov, I. B. 2012. Development trends in the production of oilseeds in the world and Russia. *Russian Agricultural Economics* 6: 85–87.

Ali, F., Yılmaz, A., Nadeem, M. A., Habyarimana, E., Subaşı, I. et al. 2019. Mobile genomic element diversity in world collection of safflower (*Carthamus tinctorius* L.) panel using iPBS-retrotransposon markers. *PLOS ONE* 14(2): e0211985. https://doi.org/10.1371/journal.pone.0211985.

Ambreen, H., Kumar, S., Variath, M. T., Joshi, G., Bali, S., Agarwal, M., Kumar, A., Jagannath, A. and Goel, S. 2015. Development of genomic microsatellite markers in *Carthamus tinctorius* L. (safflower) using next generation sequencing and assessment of their cross-species transferability and utility for diversity analysis. *PloS One* 10(8): e0135443.

Amini, F., Saeidi, G. and Arzani, A. 2008. Study of genetic diversity in safflower genotypes using agro-morphological traits and RAPD markers. *Euphytica* 163(1): 21–30.

Arslan, B. 2007. The path analysis of yield and its components in safflower (*Carthamus tinctorius* L.). *Journal of Biological Sciences* 7(4): 668–672.

Azevedo Peixoto, L. de, Laviola, B. G., Alves, A. A., Rosado, T. B. and Bhering, L. L. 2017. Breeding *Jatropha curcas* by genomic selection: A pilot assessment of the accuracy of predictive models. *PLoS One* 12(3): e0173368.

Babu, B. K., Mathur, R. K., Kumar, P. N., Ramajayam, D., Ravichandran, G., Venu, M. V. B. and Babu, S. S. 2017. Development, identification and validation of CAPS marker for SHELL trait which governs dura, pisifera and tenera fruit forms in oil palm (*Elaeis guineensis* Jacq.). *PLoS One* 12(2): e0171933.

Bai, L., Zhou, P., Li, D. and Ju, X. 2017. Changes in the gastrointestinal microbiota of children with acute lymphoblastic leukaemia and its association with antibiotics in the short term. *Journal of Medical Microbiology* 66(9): 1297–1307.

Baigot, M. and Glotova, I. 2017. Development of the oil and fat industry in the EAEU member states. *Fat-and-Oil Industry* 1(2): 12–17.

Bao, Y., Vuong, T., Meinhardt, C., Tiffin, P., Denny, R. et al. 2014. Potential of association mapping and genomic selection to explore PI 88788 derived soybean cyst nematode resistance. *The Plant Genome* 7(3), plantgenome2013.11.0039.

Baute, G. J., Owens, G. L., Bock, D. G. and Rieseberg, L. H. 2016. Genome-wide genotyping-by-sequencing data provide a high-resolution view of wild Helianthus diversity, genetic structure, and interspecies gene flow. *American Journal of Botany* 103(12): 2170–2177.

Bellaloui, N., Bruns, H. A., Abbas, H. K., Mengistu, A., Fisher, D. K. et al. 2015. Effects of row-type, row-spacing, seeding rate, soil-type, and cultivar differences on soybean seed nutrition under us Mississippi Delta conditions. *PloS One* 10(6): e0129913.

Bernardo, R. and Yu, J. 2007. Prospects for genomewide selection for quantitative traits in maize. *Crop Science* 47(3): 1082–1090.

Billotte, N., Jourjon, M.-F., Marseillac, N., Berger, A., Flori, A. et al. 2010. QTL detection by multi-parent linkage mapping in oil palm (Elaeis guineensis Jacq.). *Theoretical and Applied Genetics* 120(8): 1673–1687.

Botstein, D., White, R. L., Skolnick, M. and Davis, R. W. 1980. Construction of a genetic linkage map in man using restriction fragment length polymorphisms. *American Journal of Human Genetics* 32(3): 314.

Bowers, J. E., Pearl, S. A. and Burke, J. M. 2016. Genetic mapping of millions of SNPs in safflower (*Carthamus tinctorius* L.) via whole-genome resequencing. *G3: Genes, Genomes, Genetics* 6(7): 2203–2211.

Celik, I., Bodur, S., Frary, A. and Doganlar, S. 2016. Genome-wide SNP discovery and genetic linkage map construction in sunflower (*Helianthus annuus* L.) using a genotyping by sequencing (GBS) approach. *Molecular Breeding* 36(9): 133. https://doi.org/10.1007/s11032-016-0558-8.

Clark, S. A., Hickey, J. M., Daetwyler, H. D. and van der Werf, J. H. 2012. The importance of information on relatives for the prediction of genomic breeding values and the implications for the makeup of reference data sets in livestock breeding schemes. *Genetics Selection Evolution* 44(1): 1–9.

Cloutier, S., Niu, Z., Datla, R. and Duguid, S. 2009. Development and analysis of EST-SSRs for flax (*Linum usitatissimum* L.). *Theoretical and Applied Genetics* 119(1): 53–63.

Cros, D., Denis, M., Bouvet, J.-M. and Sanchez, L. 2015. *Genomic Selection for Heterosis Without Dominance in Multiplicative Traits: Case Study of Bunch Production in Oil Palm [W554].*

Cros, D., Bocs, S., Riou, V., Ortega-Abboud, E., Tisné, S. et al. 2017. Genomic preselection with genotyping-by-sequencing increases performance of commercial oil palm hybrid crosses. *BMC Genomics* 18(1): 1–17.

Crossa, J., Campos, G. de los, Pérez, P., Gianola, D., Burgueño, J. et al. 2010. Prediction of genetic values of quantitative traits in plant breeding using pedigree and molecular markers. *Genetics* 186(2): 713–724.

Daetwyler, H. D., Hickey, J. M., Henshall, J. M., Dominik, S., Gredler, B. et al. 2010. Accuracy of estimated genomic breeding values for wool and meat traits in a multi-breed sheep population. *Animal Production Science* 50(12): 1004–1010.

Davey, J. W., Hohenlohe, P. A., Etter, P. D., Boone, J. Q., Catchen, J. M. and Blaxter, M. L. 2011. Genome-wide genetic marker discovery and genotyping using next-generation sequencing. *Nature Reviews Genetics* 12(7): 499–510.

de Los Campos, G., Gianola, D., Rosa, G. J., Weigel, K. A., Crossa, J. et al. 2010. Semi-parametric genomic-enabled prediction of genetic values using reproducing kernel Hilbert spaces methods. *Genetics Research* 92(4): 295–308.

de Los Campos, G., Hickey, J. M., Pong-Wong, R., Daetwyler, H. D., Calus, M. P. et al. 2013. Whole-genome regression and prediction methods applied to plant and animal breeding. *Genetics* 193(2): 327–345.

Diederichsen, A., Kusters, P. M., Kessler, D., Bainas, Z., Gugel, R. K. et al. 2013. Assembling a core collection from the flax world collection maintained by Plant Gene Resources of Canada. *Genetic Resources and Crop Evolution* 60(4): 1479–1485.

Dimitrijevic, A. and Horn, R. 2018. Sunflower hybrid breeding: From markers to genomic selection. *Frontiers in Plant Science* 8: 2238.

Duhnen, A., Gras, A., Teyssèdre, S., Romestant, M., Claustres, B. et al. 2017. Genomic selection for yield and seed protein content in soybean: A study of breeding program data and assessment of prediction accuracy. *Crop Science* 57(3): 1325–1337.

Edwards, K., Johnstone, C. and Thompson, C. 1991. A simple and rapid method for the preparation of plant genomic DNA for PCR analysis. *Nucleic Acids Research* 19(6): 1349.

Edwards, S. M., Buntjer, J. B., Jackson, R., Bentley, A. R., Lage, J. et al. 2019. The effects of training population design on genomic prediction accuracy in wheat. *Theoretical and Applied Genetics* 132(7): 1943–1952.

Freitas, L. E. L., Nunes, A. J. P. and do Carmo Sá, M. V. 2011. Growth and feeding responses of the mutton snapper, Lutjanus analis (Cuvier 1828), fed on diets with soy protein concentrate in replacement of Anchovy fish meal. *Aquaculture Research* 42(6): 866–877.

Gaynor, R. C., Gorjanc, G., Bentley, A. R., Ober, E. S., Howell, P., Jackson, R., Mackay, I. J. and Hickey, J. M. 2017. A two-part strategy for using genomic selection to develop inbred lines. *Crop Science* 57(5): 2372–2386.

Goddard, M. 2009. Genomic selection: Prediction of accuracy and maximisation of long term response. *Genetica* 136(2): 245–257.

Golparvar, A. R. 2011. Assessment of relationship between seed and oil yield with agronomic traits in spring safflower cultivars under drought stress condition. *Journal of Research in Agricultural Science* 7(2(13)): 109–113.

Green, A. G. 1986. A mutant genotype of flax (*Linum usitatissimum* L.) containing very low levels of linolenic acid in its seed oil. *Canadian Journal of Plant Science* 66(3): 499–503.

Habier, D., Fernando, R. L. and Dekkers, J. C. 2007. The impact of genetic relationship information on genome-assisted breeding values. *Genetics* 177(4): 2389–2397.

Habier, D., Fernando, R. L., Kizilkaya, K. and Garrick, D. J. 2011. Extension of the Bayesian alphabet for genomic selection. *BMC Bioinformatics* 12(1): 1–12.

Hayes, B. J., Bowman, P. J., Chamberlain, A. J. and Goddard, M. E. 2009. Invited review: Genomic selection in dairy cattle: Progress and challenges. *Journal of Dairy Science* 92(2): 433–443.

Heffner, E. L., Jannink, J.-L. and Sorrells, M. E. 2011. Genomic selection accuracy using multifamily prediction models in a wheat breeding program. *The Plant Genome* 4(1).

Hickey, J. M., Dreisigacker, S., Crossa, J., Hearne, S., Babu, R. et al. 2014. Evaluation of genomic selection training population designs and genotyping strategies in plant breeding programs using simulation.

Hidalgo, A. M., Bastiaansen, J. W. M., Lopes, M. S., Veroneze, R., Groenen, M. A. M. and De Koning, D.-J. 2015. Accuracy of genomic prediction using deregressed breeding values estimated from purebred and crossbred offspring phenotypes in pigs. *Journal of Animal Science* 93(7): 3313–3321.

Howard, R., Carriquiry, A. L. and Beavis, W. D. 2014. Parametric and nonparametric statistical methods for genomic selection of traits with additive and epistatic genetic architectures. *G3: Genes, Genomes, Genetics* 4(6): 1027–1046.

Hyten, D. L., Pantalone, V. R., Sams, C. E., Saxton, A. M., Landau-Ellis, D. et al. 2004. Seed quality QTL in a prominent soybean population. *Theoretical and Applied Genetics* 109(3): 552–561.

Ithnin, M., Xu, Y., Marjuni, M., Serdari, N. M., Amiruddin, M. D. et al. 2017. Multiple locus genome-wide association studies for important economic traits of oil palm. *Tree Genetics & Genomes* 13(5): 1–14.

Jarquín, D., Kocak, K., Posadas, L., Hyma, K., Jedlicka, J. et al. 2014. Genotyping by sequencing for genomic prediction in a soybean breeding population. *BMC Genomics* 15(1): 1–10.

Jenko, J., Wiggans, G. R., Cooper, T. A., Eaglen, S. A. E., Luff, W. de L. et al. 2017. Cow genotyping strategies for genomic selection in a small dairy cattle population. *Journal of Dairy Science* 100(1): 439–452.

Jin, J., Lee, M., Bai, B., Sun, Y., Qu, J., Alfiko, Y., Lim, C. H., Suwanto, A., Sugiharti, M. and Wong, L. 2016. Draft genome sequence of an elite Dura palm and whole-genome patterns of DNA variation in oil palm. *DNA Research* 23(6): 527–533.

Khan, M. A., von Witzke-Ehbrecht, S., Maass, B. L. and Becker, H. C. 2009. Relationships among different geographical groups, agro-morphology, fatty acid composition and RAPD marker diversity in safflower (*Carthamus tinctorius*). *Genetic Resources and Crop Evolution* 56(1): 19–30.

Kumar, S., Ambreen, H., Murali, T. V., Bali, S., Agarwal, M. et al. 2015. Assessment of genetic diversity and population structure in a global reference collection of 531 accessions of *Carthamus tinctorius* L. (Safflower) using AFLP markers. *Plant Molecular Biology Reporter* 33(5): 1299–1313.

Kumari, S., Choudhary, R. C., Kumara Swamy, R. V., Saharan, V., Joshi, A. et al. 2017. Assessment of genetic diversity in safflower (*Carthamus tinctorius* L.) genotypes through morphological and SSR marker. *J. Pharmacogn. Phytochem.* 6(5): 2723–2731.

Kwong, Q. B., Teh, C. K., Ong, A. L., Heng, H. Y., Lee, H. L., Mohamed, M., Low, J. Z.-B., Apparow, S., Chew, F. T. and Mayes, S. 2016. Development and validation of a high-density SNP genotyping array for African oil palm. *Molecular Plant* 9(8): 1132–1141.

Kwong, Q. B., Ong, A. L., Teh, C. K., Chew, F. T., Tammi, M. et al. 2017. Genomic selection in commercial perennial crops: Applicability and improvement in oil palm (Elaeis guineensis Jacq.). *Scientific Reports* 7(1): 1–9.

Lee, G.-A., Sung, J.-S., Lee, S.-Y., Chung, J.-W., Yi, J.-Y. et al. 2014. Genetic assessment of safflower (*Carthamus tinctorius* L.) collection with microsatellite markers acquired via pyrosequencing method. *Molecular Ecology Resources* 14(1): 69–78.

Litt, M. and Luty, J. A. 1989. A hypervariable microsatellite revealed by *in vitro* amplification of a dinucleotide repeat within the cardiac muscle actin gene. *American Journal of Human Genetics* 44(3): 397.

Liu, G., Zhao, Y., Gowda, M., Longin, C. F. H., Reif, J. C. et al. 2016. Predicting hybrid performances for quality traits through genomic-assisted approaches in Central European wheat. *PLoS One* 11(7): e0158635.

López, F., Alfaro, A., García, M. M., Díaz, M. J., Calero, A. M. et al. 2004. Pulp and paper from tagasaste (Chamaecytisus proliferus LF ssp. Palmensis). *Chemical Engineering Research and Design* 82(8): 1029–1036.

Ma, G. J., Markell, S. G., Song, Q. J. and Qi, L. L. 2017. Genotyping-by-sequencing targeting of a novel downy mildew resistance gene Pl 20 from wild Helianthus argophyllus for sunflower (*Helianthus annuus* L.). *Theoretical and Applied Genetics* 130(7): 1519–1529.

Mangin, B., Bonnafous, F., Blanchet, N., Boniface, M.-C., Bret-Mestries, E. et al. 2017. Genomic prediction of sunflower hybrids oil content. *Frontiers in Plant Science* 8: 1633.

Marchal, A., Legarra, A., Tisne, S., Carasco-Lacombe, C., Manez, A., Suryana, E., Omoré, A., Nouy, B., Durand-Gasselin, T. and Sánchez, L. 2016. Multivariate genomic model improves analysis of oil palm (*Elaeis guineensis* Jacq.) progeny tests. *Molecular Breeding* 36(1): 2.

Meuwissen, T. H., Hayes, B. J. and Goddard, M. E. 2001. Prediction of total genetic value using genome-wide dense marker maps. *Genetics* 157(4): 1819–1829.

Meuwissen, T. H. 2009. Accuracy of breeding values of 'unrelated' individuals predicted by dense SNP genotyping. *Genetics Selection Evolution* 41(1): 1–9.

Muranty, H., Jorge, V., Bastien, C., Lepoittevin, C., Bouffier, L. et al. 2014. Potential for marker-assisted selection for forest tree breeding: Lessons from 20 years of MAS in crops. *Tree Genetics & Genomes* 10(6): 1491–1510.

Myint, K. A., Amiruddin, M. D., Rafii, M. Y., Abd Samad, M. Y., Ramlee, S. I. et al. 2019. Genetic diversity and selection criteria of MPOB-Senegal oil palm (Elaeis guineensis Jacq.) germplasm by quantitative traits. *Industrial Crops and Products* 139: 111558.

Neyhart, J. L., Tiede, T., Lorenz, A. J. and Smith, K. P. 2017. Evaluating methods of updating training data in long-term genomewide selection. *G3: Genes, Genomes, Genetics* 7(5): 1499–1510.

Nichols, D. M., Glover, K. D., Carlson, S. R., Specht, J. E., Diers, B. W. et al. 2006. Fine mapping of a seed protein QTL on soybean linkage group I and its correlated effects on agronomic traits. *Crop Science* 46(2): 834–839.

Nyouma, A., Bell, J. M., Jacob, F. and Cros, D. 2019. From mass selection to genomic selection: One century of breeding for quantitative yield components of oil palm (Elaeis guineensis Jacq.). *Tree Genetics & Genomes* 15(5): 1–16.

OECD, F. 2020. *OECD-FAO Agricultural Outlook 2020–2029*. OECD.

O'Shea, R., Lin, R., Wall, D. M., Browne, J. D., Murphy, J. D. et al. 2020. Using biogas to reduce natural gas consumption and greenhouse gas emissions at a large distillery. *Applied Energy* 279: 115812.

Pszczola, M. and Calus, M. P. L. 2016. Updating the reference population to achieve constant genomic prediction reliability across generations. *Animal* 10(6): 1018–1024.

Qi, Z., Helmers, M. J., Malone, R. W. and Thorp, K. R. 2011. Simulating long-term impacts of winter rye cover crop on hydrologic cycling and nitrogen dynamics for a corn-soybean crop system. *Transactions of the ASABE* 54(5): 1575–1588.

Rahman, N. H. A., Abd Aziz, S. and Hassan, M. A. 2013. Production of ligninolytic enzymes by newly isolated bacteria from palm oil plantation soils. *Bioresources* 8(4): 6136–6150.

Ramasubramanian, V. and Beavis, W. D. 2020. Factors Affecting Response to Recurrent Genomic Selection in Soybeans. *BioRxiv*.

Rauf, S., Jamil, N., Tariq, S. A., Khan, M., Kausar, M. et al. 2017. Progress in modification of sunflower oil to expand its industrial value. *Journal of the Science of Food and Agriculture* 97(7): 1997–2006.

Reif, J. C., Zhao, Y., Würschum, T., Gowda, M. and Hahn, V. 2013. Genomic prediction of sunflower hybrid performance. *Plant Breeding* 132(1): 107–114.

Seiler, G. J., Qi, L. L. and Marek, L. F. 2017. Utilization of sunflower crop wild relatives for cultivated sunflower improvement. *Crop Science* 57(3): 1083–1101.

Shu, Y. J., Yu, D. S., Wang, D., Bai, X., Zhu, Y. M. et al. 2013a. Genomic selection of seed weight based on low-density SCAR markers in soybean. *Genetics and Molecular Research* 12(3): 2178–2188.

Shu, Y. J., Yu, D. S., Wang, D., Bai, X., Zhu, Y. M. et al. 2013b. Genomic selection of seed weight based on low-density SCAR markers in soybean. *Genetics and Molecular Research* 12(3): 2178–2188.

Silaeva, L. P. 2016. Formation of specialized zones for the production of certain types of agricultural products, raw materials and food based on the improvement of interregional exchange. *Development of AIC in Russia: Trends and Prospects: Conference Materials at the III Moscow Economic Forum (Moscow: FGBBNU VNIIESKh)*, 220–228.

Simanjuntak, J. D., von Cramon-Taubadel, S. and Kusnadi, N. 2020. Vertical price transmission in soybean, soybean oil, and soybean meal markets. *Jurnal Manajemen & Agribisnis* 17(1): 42–42.

Singh, A., Orsat, V. and Raghavan, V. 2013. Soybean hydrophobic protein response to external electric field: A molecular modeling approach. *Biomolecules* 3(1): 168–179.

Song, Q., Jia, G., Zhu, Y., Grant, D., Nelson, R. T. et al. 2010. Abundance of SSR motifs and development of candidate polymorphic SSR markers (BARCSOYSSR_1. 0) in soybean. *Crop Science* 50(5): 1950–1960.

Soto-Cerda, B. J., Maureira-Butler, I., Muñoz, G., Rupayan, A., Cloutier, S. et al. 2012. SSR-based population structure, molecular diversity and linkage disequilibrium analysis of a collection of flax (*Linum usitatissimum* L.) varying for mucilage seed-coat content. *Molecular Breeding* 30(2): 875–888. https://doi.org/10.1007/s11032-011-9670-y.

Soto-Cerda, B. J., Duguid, S., Booker, H., Rowland, G., Diederichsen, A. et al. 2014. Association mapping of seed quality traits using the Canadian flax (*Linum usitatissimum* L.) core collection. *Theoretical and Applied Genetics* 127(4): 881–896.

Sun, Y., Pan, J., Shi, X., Du, X., Wu, Q. et al. 2012. Multi-environment mapping and meta-analysis of 100-seed weight in soybean. *Molecular Biology Reports* 39(10): 9435–9443.

Ting, N.-C., Jansen, J., Mayes, S., Massawe, F., Sambanthamurthi, R. et al. 2014. High density SNP and SSR-based genetic maps of two independent oil palm hybrids. *BMC Genomics* 15(1): 1–11.

Ting, N.-C., Mayes, S., Massawe, F., Sambanthamurthi, R., Jansen, J., Alwee, S. S. R. S. et al. 2018. Putative regulatory candidate genes for QTL linked to fruit traits in oil palm (Elaeis guineensis Jacq.). *Euphytica* 214(11): 1–16.

Tisné, S., Denis, M., Cros, D., Pomiès, V., Riou, V. et al. 2015. Mixed model approach for IBD-based QTL mapping in a complex oil palm pedigree. *BMC Genomics* 16(1): 1–12.

Varshney, R. K., Shi, C., Thudi, M., Mariac, C., Wallace, J., Qi, P., Zhang, H., Zhao, Y., Wang, X. and Rathore, A. 2017. Pearl millet genome sequence provides a resource to improve agronomic traits in arid environments. Nature Biotechnology 35(10): 969–976.

Wang, X.-Z., Harding, H. P., Zhang, Y., Jolicoeur, E. M., Kuroda, M. and Ron, D. 1998. Cloning of mammalian Ire1 reveals diversity in the ER stress responses. *The EMBO Journal* 17(19): 5708–5717.

Wang, Y., Han, Y., Teng, W., Zhao, X., Li, Y. et al. 2014. Expression quantitative trait loci infer the regulation of isoflavone accumulation in soybean (*Glycine max* L. Merr.) seed. *BMC Genomics* 15(1): 1–11.

Warrington, C. V., Abdel-Haleem, H., Hyten, D. L., Cregan, P. B., Orf, J. H., Killam, A. S. et al. 2015. QTL for seed protein and amino acids in the Benning× Danbaekkong soybean population. *Theoretical and Applied Genetics* 128(5): 839–850.

Weiss, E. A. 1983. *Oilseed Crops*. Longman Group.

Williams, J. G., Kubelik, A. R., Livak, K. J., Rafalski, J. A. and Tingey, S. V. 1990. DNA polymorphisms amplified by arbitrary primers are useful as genetic markers. *Nucleic Acids Research* 18(22): 6531–6535.

Wong, C. K. and Bernardo, R. 2008. Genomewide selection in oil palm: Increasing selection gain per unit time and cost with small populations. *Theoretical and Applied Genetics* 116(6): 815–824.

Wu, Z., Liu, H., Zhan, W., Yu, Z., Qin, E. et al. 2021. The chromosome-scale reference genome of safflower (*Carthamus tinctorius*) provides insights into linoleic acid and flavonoid biosynthesis. *Plant Biotechnology Journal* 1–18.

Würschum, T., Langer, S. M., Longin, C. F. H., Korzun, V., Akhunov, E., Ebmeyer, E., Schachschneider, R., Schacht, J., Kazman, E. and Reif, J. C. 2013. Population structure, genetic diversity and linkage disequilibrium in elite winter wheat assessed with SNP and SSR markers. *Theoretical and Applied Genetics* 126(6): 1477–1486.

Yadava, S. K., Arumugam, N., Mukhopadhyay, A., Sodhi, Y. S., Gupta, V., Pental, D. and Pradhan, A. K. 2012. QTL mapping of yield-associated traits in *Brassica juncea*: Meta-analysis and epistatic interactions using two different crosses between east European and Indian gene pool lines. *Theoretical and Applied Genetics* 125(7): 1553–1564.

Yang, Y.-X., Wu, W., Zheng, Y.-L., Chen, L., Liu, R.-J. et al. 2007. Genetic diversity and relationships among safflower (*Carthamus tinctorius* L.) analyzed by inter-simple sequence repeats (ISSRs). *Genetic Resources and Crop Evolution* 54(5): 1043–1051.

Yi, N. and Xu, S. 2008. Bayesian LASSO for quantitative trait loci mapping. *Genetics* 179(2): 1045–1055.

You, F. M., Booker, H. M., Duguid, S. D., Jia, G. and Cloutier, S. 2016. Accuracy of genomic selection in biparental populations of flax (*Linum usitatissimum* L.). *The Crop Journal* 4(4): 290–303.

You, F. M., Cloutier, S., Rashid, K. Y. and Duguid, S. D. 2019. Flax (*Linum usitatissimum* L.) genomics and breeding. pp. 277–317. *In*: *Advances in Plant Breeding Strategies: Industrial and Food Crops*. Springer.

Zhang, J., Song, Q., Cregan, P. B. and Jiang, G.-L. 2016. Genome-wide association study, genomic prediction and marker-assisted selection for seed weight in soybean (*Glycine max*). *Theoretical and Applied Genetics* 129(1): 117–130.

Zhang, A., Wang, H., Beyene, Y., Semagn, K., Liu, Y. et al. 2017. Effect of trait heritability, training population size and marker density on genomic prediction accuracy estimation in 22 bi-parental tropical maize populations. *Frontiers in Plant Science* 8: 1916.

Zhao, H., Li, Y., Petkowski, J., Kant, S., Hayden, M. J. et al. 2021. Genomic prediction and genomic heritability of grain yield and its related traits in a safflower genebank collection. *The Plant Genome* 14(1): e20064.

Zhao, Y., Zeng, J., Fernando, R. and Reif, J. C. 2013. Genomic prediction of hybrid wheat performance. *Crop Science* 53(3): 802–810.

Zhao, Y., Mette, M. F. and Reif, J. C. 2015. Genomic selection in hybrid breeding. *Plant Breeding* 134(1): 1–10.

Zhao-ming, Q., Ya-nan, S., Qiong, W., Chun-yan, L., Guo-hua, H. et al. 2011. A meta-analysis of seed protein concentration QTL in soybean. *Canadian Journal of Plant Science* 91(1): 221–230.

9

Harnessing the Potential of Genomic Selection in Potato

Praveen Kumar Oraon,[1] *Priyanka*[1] and *Sapinder Bali*[2,*]

ABSTRACT

Potato is the third most important food crop in the world with high nutritive value for feeding the growing world population and economic value for the growers. Conventional potato breeding for trait improvement is a slow and tedious process with added complications due to its complex genetic makeup including varying ploidy levels, heterozygosity, intra-species crossing incompatibilities and inbreeding depression. Genomic selection (GS) is becoming increasingly applicable in potato as high-throughput genotyping and computational methods advance; and the costs continue to decrease. With the availability of genome-wide molecular markers and predictions entirely based on training populations in the absence of direct phenotyping, crop improvement programs using GS in potato will result in more gains per unit time. Thus, GS promises a better approach for precise and reliable prediction and shortening the long and tedious breeding cycle. This chapter overall addresses the road to genomic selection in potato and its usefulness in potato breeding programs.

Introduction

The world's population is currently growing at a rate of around 1.05% per year. The average population increase is estimated at around 81 million per

[1] Department of Botany, University of Delhi, New Delhi, India.
[2] Department of Plant Pathology, Washington State University, Washington, USA.
* Corresponding author: sapinder24jan@gmail.com

year and the total population will reach 10.9 billion by 2100 (World Population Prospects, 2019: Ten Key Findings). The continuous growth in population has led to a greater demand for food security (Godfray et al., 2010). Among the other factors that affect the attainment of food security, insufficient food production is a major limitation (Premanandh, 2011). Attaining food security would mean growing more food, hence competition for land, water, and energy apart from the overexploitation of fisheries, sea foods, and foods available from the forest (Von Braun, 2007). Food shortage could only be met by changing the way food is produced, stored, processed, distributed, and by preferably growing crops with higher yields and nutritive values, and resilience to climate change.

Based on human consumption, potato ranks third after rice and wheat and fifth based on its total production worldwide (FAOSTAT, 2018). Potato cultivation has steadily expanded globally, with a 21% increase in overall production over the past two decades, indicating its importance as a staple food source. However, the area harvested per hectare has continuously decreased from 2000 to 2019 as compared to the increase in production per hectare (Fig. 1).

Potato, *Solanum tuberosum*, is an auto-tetraploid species (2n = 4× = 48) first introduced to Europe and North America, and later spread as a botanical novelty and as fodder crop for livestock to the rest of the world (Kumari et al., 2018). Autopolyploidization of diploid *S. tuberosum* groups Stenotomum and Phureja occurred repeatedly, and this process resulted in Andean cultivated tetraploid, *S. tuberosum* group Andigena (2n = 4x = 48). Migration of potato from Andes to coastal Chile produced a long day adapted subspecific group Chilotanum (2n = 4x = 48), which is significantly distinct from its progenitors.

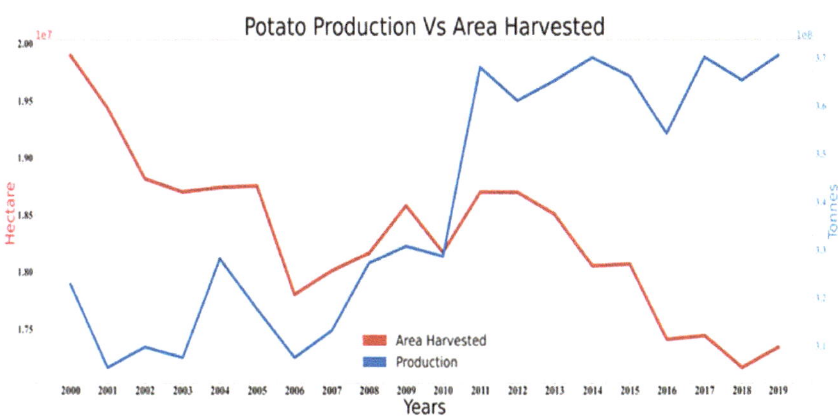

Fig. 1. Line graph showing potato production (in tons) versus area harvested (in hectare) from year 2000–2019 (FAOSTAT, 2018).

Subsequently this group contributed to most of the genetic background in commercial cultivars of *S. tuberosum*.

Conventional potato breeding has been practiced for over hundred years resulting in many new commercially acceptable varieties. Potato breeding majorly involves selection of parents with traits of interest, followed by traditional controlled crosses. Selection of superior breeding lines is performed by germinating true potato seeds (TPS) resulting from the controlled crosses as a starting material and rigorous phenotypic selections thereafter. Phenotypic recurrent selection is continued for several successive years from nonreplicated to replicated single hill selections and finally to multistate replicated trials. Seed size of the selected progenies increases due to clonal propagation whereas the number of selections reduces significantly, with only 10% making it to the replicated multistate trials (Jansky et al., 2018). Although potato breeding is a long and tedious process, takes ~ 12–14 years from the first cross to an improved and acceptable variety release hundreds of commercial potato varieties have been developed and released since the mid-1800s (Jansky et al., 2018). The gene pool for potato breeding is limited due to its genetic factors like, variable ploidy levels (2n–6n) in section *Petota*, and most wild species being diploids resulting in crossing barriers. Hence, a majority of the genetic diversity in Solanum section *Petota* is largely unexplored (Hardigan et al., 2017). However, to overcome these barriers, ploidy manipulation, somatic fusions, somatic doubling, and vegetative propagation have played a substantial role in the introgression of desirable traits from wild to cultivated potatoes (Bethke et al., 2017). Successful interspecific crosses rely mainly on endosperm balance number (EBN) although interspecific hybrids are known to show cytoplasmic male sterility and self-incompatibility which adds to the restricted use of wild species in potato breeding. In addition, the tetraploid nature of the cultivated potato increases its genetic complexity and therefore, hinders the introgression of new traits and sometimes unfavorable alleles can pass unnoticed through generations. The highly heterozygous nature of potato presents additional limitations in successful breeding, with only one meiotic event during the crossing step that limits more recombination events and rearrangements of allelic combinations. Due to such genome complexities, breeding barriers and longer breeding cycles breeders have failed to achieve major accomplishments in potatoes in the past 150 years as compared to other crops (Jansky et al., 2009). Longer breeding cycles make it harder to maintain pathogen free seeds for successive generations as potato is clonally propagated and pathogens such as seed born viruses accumulate in successive generations (Priegnitz et al., 2019). Pathogens could also alter or reduce the fitness of the infected clones, resulting in poor performance hence, voiding the selection step, which is typically based on phenotypic evaluations in conventional breeding (Jansky et al., 2018). Breeders could use self-compatible wild diploid relatives and dihaploids of cultivated tetraploid potatoes to facilitate potato breeding programs, but the challenges still persist.

The limitations of conventional breeding could be addressed by marker assisted breeding (MAB) that involves the use of markers to select the traits of interest hence replacing the tedious and time-consuming phenotypic evaluations. Different marker techniques have been developed over the years to measure and exploit genetic variation present within and between plant breeding populations essential for the selection of superior genotypes (Clegg et al., 1990). These include conventional classical markers (based mostly on morphological or phenotypic characters), cytological and biochemical molecules (like isozymes) and the molecular markers (DNA-based markers). Classical markers are limited in number and are affected by the developmental stage of the plant and environment whereas, molecular markers are present in abundance in the genome and are relatively more stable. Marker assisted selection (MAS) and genomic selection (GS) are the commonly used tools in MAB (Bhat et al., 2016). In MAS, molecular markers associated with the trait(s) of interest are used for the selection of desired phenotypes all through the breeding cycle. Marker systems have been successfully utilized to create linkage maps for identifying, and localizing markers closely linked with quantitative traits of interest thus, allowing researchers to quantify and select genetic components of QTLs (Edwards et al., 1987; Paterson et al., 1988). In the past, the majority of marker-based linkage mappings were restricted to diploid wild relatives or diploid lines derived from cultivated potato because the heterozygous autotetraploid nature and tetrasomic inheritance restricted their direct use in cultivated potatoes (Barrell et al., 2013). The first linkage map of diploid potatoes was constructed using RFLP markers reported in tomatoes (Bonierbale et al., 1988). Marker-based mapping has led to the discovery of twenty single dominant genes linked to resistance to various viruses (Ritter et al., 1991), nematodes (Barone et al., 1990; Gebhardt et al., 1993; Jacobs et al., 1996; Van der Voort et al., 1997; Brown et al., 1996) and fungi (Naess et al., 2000; Kuhl et al., 2001; Hehl et al., 1999). Recent advancements in computer programing have made it possible to develop genetic maps using autotetraploid potato populations (Hackett et al., 2003; Moloney et al., 2010). The first linkage map for a tetraploid potato was developed for the F1population (Stirling X 12601ab1-an advanced SCRI clone) using AFLP and SSR markers with the mapped markers condensed mostly around the trait of interest (Meyer et al., 1998). MAS of quantitative traits based on pedigree information further reduces the number of breeding cycles by 4–5 years and therefore results in more genetic gain per unit time compared to conventional breeding (Slater et al., 2014a, b). However, selections based on QTL mapping performed using segregation of biparental populations presents several limitations such as low resolution genetic maps, strong linkage disequilibrium (LD), difficulty in detecting complex QTLs with minor effects and identification of the same QTLs across different environments (G × E interaction) or in different genetic backgrounds (Parisseaux and Bernardo, 2004) and to circumvent these shortcomings, linkage disequilibrium-based genome-wide association studies (GWAS) or association mapping (AM) was introduced.

GWAS uses panels of unrelated genotypes for creating highly resolved genetic maps by utilizing natural genetic diversity of a species and its historical recombinations (Gupta et al., 2014). With the availability of efficient and affordable, high-throughput, genome wide markers, GWAS offers new potential for characterizing traits of interest in both plants and animals thus, exploiting high allelic diversity for QTL detection (Neumann et al., 2011). In potato, GWAS has helped in identifying QTLs for various important traits such as resistance to pathogens, starch content, chip quality, tuber yield and foliage maturity (Reviewed by Ortiz, 2020). The first GWAS study on potatoes was performed in 2004 (Gebhardt et al., 2004) followed by several more studies for detecting various traits of interest (Simko et al., 2004; Li et al., 2008; Pajerowska-Mukhtar et al., 2009; Fischer et al., 2013; Lindqvist-Kreuze et al., 2014; Mosquera et al., 2016; Juyo Rojas et al., 2019). GWAS helps dissect the genetic basis of complex traits and offers an advantage to localize QTLs in a panel of diverse individuals for high resolution mapping of candidate genes (Michel et al., 2019) but the estimation of cumulative QTL effects of complex traits using it, is often biased leading to incorrect genotypic selections based solely on markers (Bhat et al., 2016). Although GWAS presents several limitations such as its inability to detect rare alleles and the effects of population structure but it also offers the ability to detect robust useful markers tagging major QTLs (Jannink, 2007). QTL identification in polyploids like potatoes is also hindered due to the presence of a larger number of possible genotypes, which determines the allele dosage (Wilson et al., 2021). However, the allele dosage problem could be overcome by directly recording the allele frequencies rather than performing discrete genotype calling (Endelman et al., 2018; de Bem Oliveira et al., 2019).

In contrast to GWAS, genome-wide prediction (GP) for genomic selection (GS) utilizes genetic markers covering the whole genome to generate a model of both genotyping and phenotyping datasets or training population (TP) to predict genomic-estimated breeding values (GEBVs) for each genotype of breeding population (BP) using their genotype scores. GEBVs can foretell the potential of an unphenotyped new cultivar, therefore bypassing the need for field testing. GS uses whole-genome regression modeling from high-throughput markers to predict both major as well as minor effect genomic regions (Meuwissen et al., 2001) and helps in selecting superior genotypes suitable either as a parent for crossing or successive breeding cycles

(K Srivastava et al., 2020). Several models have been developed for estimating genomic prediction for GS but genomic best linear unbiased predictor (GBLUP), a mixed model, is the most commonly used. This model uses the relationship between individuals as input and a ridge regression penalty with a presumed normal distribution for marker effects (Piepho, 2009). The resulting relationship matrix can be derived with an assumption of both additive and non-additive effects (dominance and epistasis). Therefore, it successfully captures total additive genetic variance of the trait and is based on all favorable alleles from parents for the overall improvement of the population. GS can also be effective in selecting low heritability traits controlled by multiple and additive minor effect QTLs (Bentley et al., 2014) and GS modeling is promising for gene pyramiding as well (Bernardo and Yu, 2007). Subsequently, a test set of the BP is derived based on the genotype information, thus reducing the scale of phenotyping requirements, and shortening the overall length of the breeding cycle (Habier et al., 2013). Since, GEBV's are based on prediction of the individual marker profiles that are genetically similar to superior profiles in the TP, they are expected to perform better in specific environmental conditions.

In polyploids, the expression of target traits is expected to be controlled by allele dosage, directly affecting phenotypes and resulting either in simple or complex effects (Osborn et al., 2003; de Bem Oliveira et al., 2019). GS has shown the inclusion of allele dosage information, thus increasing the accuracy of results in polyploids (de Bem Oliveira et al., 2019). With the development of cost-effective sequencing technologies and the advancements in statistical modeling, it is now possible to sequence and utilize complex polyploid genomes for GS based breeding approaches (Garcia et al., 2013). Various methods that are being used to genotype the polyploid allelic dosage information include, high-throughput SNP genotyping (Gidskehaug et al., 2011) and genotyping by sequencing (GBS) (Ashraf et al., 2014). High-throughput SNP genotyping has relevance for the crops where SNP arrays developed from one cultivar with high transferability are readily available for inexpensive genotyping studies. In contrast GBS could be used to discover genetic variants (SNP markers) between two or more diverse or related cultivars and subsequently used for genotyping larger related panels. It uses restriction digestion of the whole genome to reduce its complexity and sequencing of smaller fragments. GBS data analysis software can vary depending upon the ploidy level and repeat content of the sequenced genome. It allows the accessibility of the relative abundance of each allele (Gerard et al., 2018) hence, magnifying its role in the GS of complex traits. Also, GS models will be very useful in a crop like inbred potato for predicting parental values using larger genomic datasets taking advantage of identity by descent. Therefore, GS shows promise in accelerating potato breeding in a more efficient and timely manner (Bachem et al., 2019).

Genomic selection in potato

Genomic selection has been successfully implemented in several animal and plant improvement programs in the past few decades by utilizing the markers in linkage disequilibrium with different traits of interest. Potato is highly heterogenous and autotetraploid due to which the pattern of the inheritance of different traits is more complicated as compared to diploids and allotetraploid crops (Dufresne et al., 2014); hence implementation of GS in potato is limited. With the advent of new sequencing technologies that led to the discovery of high-throughput SNP markers in potato (Hamilton et al., 2011; Uitdewilligen et al., 2013; Vos et al., 2015) and development of more sophisticated data analysis programs, GS has been successfully applied in potatoes (Habyarimana et al., 2017; Sverrisdottir et al., 2017; Enciso-Rodriguez et al., 2018; Endelman et al., 2018; Amadeu et al., 2020; Ortiz, 2020; Byrne et al., 2020). The major target traits for GS in potato breeding include tuber traits such as weight, shape and number, tuber quality defined by starch content and reducing sugars, host plant resistance to pathogens and pests and other traits of importance to the producers and stakeholders. Many of these traits are determined by a single gene effect, whereas others are influenced by multiple genetic factors (Slater et al., 2014a).

In 2017, Ephrem Habyarimana et al. evaluated the potential of GS in predicting total yield and its components in tetraploid potato using DArT (Diversity Array Technology) markers. They fingerprinted 190 potato clones with > 78K markers considering at least 11 traits of interest. This study addressed two important potato breeding rationales that could be addressed by GS, including efficient prediction of performance of unphenotyped clones using 10-fold cross validation and alternate prediction of performance of a testing set using another training set. Notably, larger the TP better will be accuracy in predicting performance. They evaluated the prediction potential of Bayesian least absolute shrinkage and selection operator (BL), genomic best linear unbiased prediction (GBLUP) using a genomic relationship matrix (parametric linear models), reproducing kernel Hilbert spaces (RKHS) using a Bayesian generalized linear regression (BGLR) (semiparametric nonlinear model), Bayes A, Bayes B and Bayes C. GBLUP, BL and RKHS performed well in their preliminary tests and hence were used for genomic predictions. Hence, GS showed potential in potato breeding and higher genetic gains per unit of time and cost for the evaluated traits.

In 2018, Enciso-Rodriguez and team studied the GS for late blight, common swab resistance in potato and also the variants contributing to it using whole genome regression methods. Phenotypic field evaluations for scab (data for nine years) and late blight (data for seven years) as collected and a total of 4,110 SNP markers were generated for the training and testing populations. Prediction accuracy estimation was performed by using two cross-validation schemes. One scheme used a fivefold cross-validation by assigning genotypes to folds by allotting the phenotypic record of a genotype to the training or testing population. Therefore, the prediction accuracy was implemented to the unphenotyped genotypes based on the training population. The genotype assignments were completely random and cross validation was repeated 100 times. In the second scheme, years were assigned to folds that resulted in an estimation of the prediction accuracy by anticipating the future performance of a genotype based on the past data (multiple years). Overall, the genomic prediction performed slightly better than the predictions based on past phenotypic data only. For estimating marker effects, prediction accuracy and variance components analyses were performed using the whole genome regression method. The main effects of a genotype were estimated using four models, genotype effect, additive (A), additive + dominance (A+D) and general (G). The genotype effect was used as a baseline, modeled without the genotype information whereas the other specifications included the genotype data. Marker effects were statistically analyzed using Bayes ridge regression (BRR) (assuming effects derived from a normal distribution) and Bayes B (assuming non-linear distribution) in the BGLR-R package, resulting in 30% and 10% genetic variance for late blight and common scab, respectively. The additive model estimated more than 90% of the genetic variance for both the traits compared to the A+D model, where inclusion of dominance diminished the additive variance explained by the allele substitution effect in model A. However, the A+D model accurately predicted the variance for late blight during cross validation. The SNP variance showed chromosomes V and IX contributing to late blight resistance and notably these chromosomes are already known to harbor resistance loci against the pathogen. In addition, a novel locus was reported on chromosome IX for common swab resistance and interestingly, this locus is in a transcription factor, WRKY, which has a role in systemic defense signaling.

Later Sverrisdóttir et al., 2018 performed genomic prediction for two important tuber traits, dry matter content and chipping quality, using two different testing population sets and also included previous training populations to see whether different populations can be combined to achieve increased prediction accuracy and more genetic gain for a trait. GBS was used to genotype three different panels of cultivars generating 167,637 SNP markers and GEBVs were estimated using GBLUP. Cross-validation of predictions suggested that the prediction accuracies were somewhat similar within each panel as well as in the panel combinations. However, prediction of chipping quality across and within populations for one of the panels was not as good as compared to that of dry matter content. They concluded that prediction accuracy was not affected by the number of markers and predictions across populations are somewhat unreliable but individual prediction models within populations can be combined to achieve more accurate genomic predictions in the tetraploid potato.

More recently, Byrne et al., 2020 used chip fry color phenotypic data (for three years) for a larger set of potato lines in combination with genotypic data (46,406 SNPs) for genomic prediction. Genotyping data was developed using GWAS and BLUP, BayesA, BL and Random Forest models were used for evaluating prediction accuracy. A major QTL on the potato chromosome X showed association with fry color with moderate accuracy. Prediction accuracies of all the different algorithms applied in the study, yielded similar results except for the Random Forest model. When few handpicked SNP markers from GWAS were used for prediction, it resulted in a lower correlation and a larger bias. Surprisingly when the Random Forest model was applied for variable selection, predicting the ability of selected SNP markers was higher compared to that using randomly selected ones but the difference faded away as the number of markers was increased. It also resulted in the identification of SNPs on other chromosomes responsible for chip processing characteristics.

These studies demonstrated a significant genetic gain by implementing GS in potato regardless of autotetraploidy, tetrasomic inheritance and inbreeding depression. As mentioned earlier, phenotypic prediction using genomic selection in potato can be affected by marker allele dosage. In order to tackle allele dosage problem, two assumptions are made: (1) There is an equal effect on the genotype by all the heterozygous genotypes known as the pseudodiploid model and (2) Each genotype has its own effect (Slater et al., 2016). The first assumption accounts for additive marker effects while the second accounts for both additive and non-additive marker effects. A matrix developed on the basis of both these models can be directly applied in GS methods like BLUP (normal distribution) and BayesB, BayesA, BayesR (Non-Linear distribution). GEBV calculations can be performed once the best fitted model is determined.

Implementation of GS in potato breeding allows more accurate prediction and selection of traits at an early clonal generation stage therefore, significantly shortening the selection phase required for evaluating a larger number of clones and hence reducing resource utilization (labor, financial cost, time) required during the process of conventional breeding. GS results in efficient and superior selections based on both phenotypic and genetic attributions and increases genetic gain per unit time. Therefore, GS will not only result in genomic predictions with higher accuracy but also shortens the breeding cycle. However, several other factors can affect GS in potato as described briefly in the following section.

Factors affecting genomic selection in potato

The recent advances in sequencing technology have allowed high throughput genotyping of larger training and testing populations, accomplishing the prerequisite for GS. However, as the GS studies are being carried out, factors affecting the accuracy of the process are being identified, which need to be addressed while performing genomic prediction and GS.

Population size and marker density

Population size and marker density directly impact the prediction accuracy. The prediction accuracy increases with an increase in population size and marker density separately/independently but the genetic gain plateaus after a certain limit (Combs and Bernardo, 2013) and this can be attributed to the genome complexity and diversity among populations (Liu et al., 2018). The population size impacts GP regardless of the number of markers; relatively smaller populations genotyped using a limited number of handpicked or randomly selected markers (Sverrisdóttir et al., 2018) do not warrant an improvement in the prediction accuracy. GS prediction requires well spread markers throughout the genome for better predictability of models, thus GBS and SNP arrays are the preferred techniques. However, both GBS and SNP arrays include a large fraction of missing data, which could impact the data stability (Darrier et al., 2019). Imputation of missing marker data showed an increase in genetic gain and accuracy of GS at low depth sequencing via GBS (Wang et al., 2020). In tetraploid ryegrass, *Lolium perenne* L., prediction accuracy was improved when 80–100K SNPs were used and using SNPs with sequencing depth between 10 and 20 substantially increased the prediction ability (Guo et al., 2018).

Both population size and marker density are interdependent and an increase in the number of markers results in an increased genetic merit in any given population size but after a certain point this factor no longer impacts the outcome of GP. Population size must also increase for accurate predictions (Muir et al., 2007; Lorenz et al., 2011) however, inclusion of a larger number of markers for prediction leads to lower precision in the outcome of GP models as there is a saturation point after which there is no effect/marginal reduction in GP (Arruda et al., 2015). Thus, evenly spread markers covering the whole genome and in linkage disequilibrium with QTLs warrants better predictions for effective GS (Lorenzana and Bernardo, 2009; Heffner et al., 2011; Arruda et al., 2015; Slater et al., 2016). In order to effectively take advantage of a larger population size, genotyping of parental lines should be performed at a high depth to achieve high density SNP markers and progeny at a low depth followed by imputation methods for imputing missing marker data. Imputation methods need to be considered carefully, as they have been reported to have a bias in predicting GBEVs (Pimentel et al., 2015; Wang et al., 2020).

Heritability

Heritability of traits also determine GP and selection accuracy. Complex traits with low heritability require relatively higher density of markers and larger population sizes, whereas traits with high heritability require significantly smaller number of markers and population size (Zhang et al., 2017). Similarly, heritability of traits of interest will govern the required size of the training population too. For a complex trait like chipping quality in potato where heritability is governed by environmental interactions, using smaller testing population sizes resulted in low genomic prediction accuracy (Sverrisdóttir et al., 2018).

Ploidy

Higher ploidy in potato increases its genome complexity, thus impacting the GS outcomes as the allele dosage limits prediction ability. Genotyping for GS needs dosage information of marker alleles. GBS and SNP fingerprinting techniques allow inclusion of allele dosage information (Uitdewilligen et al., 2013; Ashraf et al., 2014; Gidskehaug et al., 2011) resulting in higher genetic merit and better GP accuracy. Exclusion of allele dosage information from GS resulted in low prediction accuracy (0.13 on average) in potatoes (Endelman et al., 2018). A study in autotetraploid perennial blueberry showed improved prediction of various traits when estimation of allele dosage from the genotypic data was included in prediction and GS (de Bem Oliveira et al., 2019). GEBVs are calculated using several methods, therefore it is also affected by the relationship matrix used (Eding and Meuwissen, 2001; VanRaden, 2008; VanRaden et al., 2011). Variation in prediction accuracy due to the relationship matrices can be caused by the information used to build the matrices, like pedigree or markers (genotype) from the sequencing data and ploidy levels. Matrices of SNP marker data are coded differently and modeled in various ways to account for additive and non-additive effects. The effect of the relationship matrix in autotetraploid blueberry revealed that the marker-based relationship matrix performed well and resulted in an increase of 11% and 13.37% in prediction accuracy compared to pedigree-based matrix (de Bem Oliveira et al., 2019). Ploidy levels not only effect Genomic Relationship Matrices (GRM) but also the genotype misclassification during genotype calling via NGS data. Addressing these factors by avoiding ploidy level assumptions showed better prediction accuracy for a few traits and reduced the timeline by surpassing complex steps of genotype calling and its parameterization (de Bem Oliveira et al., 2019). Minor allele frequency

(MAF) is a critical parameter in SNP filtering as the genomic relationship matrix uses genotype data and a low MAF showed poor prediction accuracy with the BayesR method (Zhang et al., 2019). However, using bins of MAF markers filtered at different parameters (0.01–0.1, 0.1–0.2, 0.2–0.3, 0.3–0.4, 0.4–0.5) revealed a higher prediction accuracy in cattle (Bo et al., 2017).

Statistical models

The performance of different prediction methods directly affects prediction accuracy therefore different models could be chosen based on certain criterion for example, Bayes models should be considered when the training and testing populations are genetically distant and the population size is relatively smaller (Fernando et al., 2007; Habier et al., 2013; Onogi et al., 2016). However various studies on model comparison used for autotetraploid species including potato show an almost similar performance with marginal differences in prediction accuracy (Stich and Van Inghelandt, 2018). However, Bayes models (Bayes A, Bayes B, Bayes C) outperformed as compared to other models (GBLUP, RKHS, BGLR) (Habyarimana et al., 2017) but similar prediction outcomes for different models like, GBLUP, BayesA, Bayes C, rrBLUP and BL have also been reported in potatoes (Sverrisdóttir et al., 2017; Byrne et al., 2020). In addition, the additive and non-additive effects that capture genetic variance of a trait of interest are also shown to affect the prediction accuracy and inclusion of various non-additive effects showed an improvement in the prediction accuracy (Endelman et al., 2018). Additive (A) model captures significant genetic variance (> 90%) and results in higher prediction accuracies as compared to the A+D model but traits governing plant fitness are under directional selection and are mainly under dominance and epistatic effects therefore, resistance to late blight trait in potato performed statistically significantly with slightly improved prediction accuracy using model A (Enciso-Rodriguez et al., 2018). Similarly, for more complex traits such as tuber starch and tuber yield, there was an 8% increase in prediction accuracy when the A+D model was used as compared to only potato model A (Stich and Van Inghelandt, 2018).

All these factors have a significant effect on prediction accuracy and GS in autotetraploid potatoes and careful implementation of GS in potato breeding will allow more efficient gain, higher prediction accuracy and higher genetic merit per unit time.

Genetic and genomic resources in potatoes

Highly diverse germplasms, both domesticated and wild, is a primary requisite for the success of any breeding program. Potato being an autotetraploid, with incompatibility issues in crossing has resulted in a narrow genetic base, hence conserving and maintaining the germplasm in order to tackle future problems is necessary. One advantage of *in situ* conservation is that it allows the maintenance of evolving populations in their natural habitats, permitting the preservation of gene frequencies and generation of genetic variability during the dynamic and permanent interaction of target populations with biotic and abiotic factors (Marfil et al., 2015). Wild potato germplasm carries useful genes for biotic as well as abiotic stress, which are absent in cultivated potato, thus increasing the need for its conservation. It also possesses high genetic diversity but is largely unexplored because of the genetic complexity and breeding barriers due to ploidy differences, thus hindering introgression of desirable traits into the cultivated potato. However, some compatible wild potatoes could easily serve the purpose of transferring traits of interest to modern cultivars by protoplast fusion (Chen et al., 2008). Gene banks around the world collect, classify, evaluate, maintain, conserve and distribute wild potato relatives along with landraces, cultivated potatoes and breeding lines hence, making them accessible to researchers especially potato breeders. According to the Consultative Group on International Agricultural Research (CGIAR) gene bank and the International Potato Centre (CIP) annual report, the potato germplasm is composed of 155 wild relatives and approximately 3000 cultivated landraces, of which Peru holds the germplasms of 80 wild potato species (CIP, 2019). According to Food and Agriculture Organization Corporate Statistical Database 2010 (http://www.fao.org/3/i1500e/i1500e.pdf) 98,285 accessions of potato are conserved and maintained *ex situ* at various institutions around the world. The comprehensive treatment of wild potato along with cultivated potato for conservation resulted in an increase of wild species collection with time, for example the number of conserved wild species increased from 10 (Walpers, 1844) to 159 (Hawkes, 1963) in a decade.

In 1990, Hawkes documented that the gene bank collections are largely composed of seven species but later Spooner et al., 2007 reclassified the database using SSR and chloroplast markers on 742 landraces into four major species viz, *Solanum tuberosum* (4n), *S. ajanhuiri* (2n), *S. juzepczukii* (3n) and *S. curtilobum* (5n). However, major changes from Hawkes to Spooner classification were the placement of three more species *S. chaucha, S. phureja, S. stenotomum* to the *S. tuberosum* subgroup, Andigenum. Major genebanks around the world such as CIP, Peru; the Leibniz Institute of Plant Genetics and Crop Plant Research, Germany; Centre for Genetic Resources, Netherlands; Germplasm Resources Information Network, USA follow Hawkes's descriptions and classifications for cataloguing and distribution of potato genetic material (Hawkes, 1990). However, Vavilov Institute of Plant Industry, Russia still follows the Spooner classification for assessing their potato collection (Gavrilenko et al., 2010).

CIP, Peru recorded their potato germplasm holdings based on 74 morphological descriptors (http://genebank.cipotato.org/gringlobal/search.aspx) as listed in the CIP web searchable database, whereas the United States Department of Agriculture-Agricultural Research Service database contains a list of 141 descriptors based on morphology, abiotic and biotic stress, physiology, cytology, genetic stock, biochemical, molecular and several other categories (https://npgsweb.ars-grin.gov/gringlobal/cropdetail.aspx?type=descriptor&id=73). Morphology based characterization in gene banks is a critical issue in case the user wants to select an accession based on the trait of interest as descriptions in gene banks vary in terms of descriptor lists. Therefore, there is an urgent need for updating potato classification and re-evaluation of germplasms to fill the information gap and avoid confusion while cross-referencing the same accessions. In order to create user-friendly and more accurate potato germplasm collections, Genesys (https://www.genesys-pgr.org/) is trying to merge all the database information onto a single platform, however it is still under progress.

Genomic resources for potatoes drastically increased with the advent of next generation sequencing (NGS) technology, hence enhancing the crop improvement programs. The first potato genome was sequenced in 2011 using Illumina and 454 sequences of BAC libraries of homozygote diploid potato (doubled monoploid derived from the anther culture) (Hein et al., 2007; Huang et al., 2005; Song et al., 2003; Vander Vossen et al., 2000) and reads were assembled into 12 pseudomolecules covering 86% (~ 726 Mb) of 844 Mb total genome size (Potato Genome Sequencing Consortium, 2011). Spud DB (http://solanaceae.plantbiology.msu.edu/), a potato genome database holds improved genomic resources of International Potato Genome Sequencing Consortium (PGSC) DM1-3 and the dyrad repository which holds *S. chacoense* M6 genomic data. This data repository is hosted and maintained by Michigan State University, USA. Subsequently, wild diploid potato *S. commersonii*, diploid inbred clone *S. chacoense* (M6) and twelve other potato landraces (ploidy ranging from diploid, triploid, tetraploid to pentaploid) were also sequenced (Kyriakidou et al., 2020).

Comparative genomics between available reference genome of DM potato and M6 diploid inbred *S. chacoense* showed that a higher number of genes were duplicated than deleted (Kyriakidou et al., 2020). Pham et al., 2017 described the sequence and structural variation in the genomes of six elite potato cultivars representing three market classes and their impact on the transcriptome. Their data suggested that in addition to allelic variation of the coding sequence, the gene expression is also impacted by the heterogenous nature of tetraploid potato that significantly impacts the transcriptome by preferential alleles and their copy number. With recent advances in sequencing technology and introduction of long read sequencing platforms like, PacBio

(Pacific Biosciences, CA, USA), Oxford Nanopore Technologies (Oxford Nanopore Technologies, Oxford, UK) and GemCode (10X Genomics, CA, USA), six potato landraces representing three ploidy levels (3x, 4x, 5x) and five taxa (*S. chaucha*, *S. juzepczukii*, *S. tuberosum* subsp. *Andigena*, *S. tuberosum* subsp. *tuberosum* and *S. curtilobum*) were sequenced showing the potential use of long reads in polyploid research (Kyriakidou et al., 2020). Recently homozygote diploid potatoDM1–3 516 R44, clone of *S. tuberosum* was re-sequenced using Oxford Nanopore Technologies (company profile) using long reads and Hi-C scaffolding to generate a high quality and improved contiguity of DM (v6.1) compared to DM1-3 (Pham et al., 2020).

Gold standard reference genomes produced via genome sequencing can be directly used in the identification of high-throughput SNP markers (i.e., genotypic data), a prerequisite for GS. Approximately 68.9 million SNPs have been identified in potatoes by resequencing sixty-seven wild and cultivated varieties (diploids and tetraploids) showing an enormous genetic diversity compared to other major crops (Hardigan et al., 2017). In addition, genome sequencing of 201 accessions including (clade 1+2, clade 3 and clade 4) representing almost the entire diploid species of *Solanum* section *Petota* resulted in the identification of 64,87,006 SNPs showing that the wild potato holds far more genetic diversity compared to modern world potatoes (Li et al., 2018). This study also showed that cultivated potato lost 7 R-genes responsible for resistance to late blight and a total of 569 other genes during its domestication when compared to *S. candolleanum* progenitor (Li et al., 2018). Therefore, wild potato species are a great source of agronomically important traits and from a breeder's perspective, diversity at specific loci is more important rather than genome wide genetic diversity. Apart from SNPs, solArray - Potato Microarray Database (https://ics.hutton.ac.uk/solarray/) retains the results of microarray experiments conducted on potatoes treated with several growth hormones (abscisic acid, aminocyclopropane carboxylic acid, epibrassinolide, methyljasmonate, and salicylic acid). PoMaMo Database - Potato Maps and More (https://www.gabipd.org/projects/Pomamo/) hosts all the potato genetic/molecular maps, SNPs, Indels, sequence data from diploid and tetraploid genotypes developed via GABI (Genomanalyse im biologischen System Pflanze).

Free accessibility of sequencing data from wild potato relatives/species, landraces and cultivated varieties representing different taxa covering the genetic diversity of potatoes and other genomic resources will allow researchers to understand interspecific genomic variations and exploit the valuable information for agronomically important traits such as resistance to various biotic and abiotic stresses for more efficient potato breeding.

Conclusion

Potato is a major contributor to the food security of a growing global population. It is an important crop and has received a lot of attention with regard to genetic improvement. There are several factors that affect its production including its own biological characteristics such as lower tuber yield and high susceptibility to soil and seed-borne insect pests and diseases. The development of new plant varieties through breeding is a powerful approach to increase both the quantity and quality of food produced per unit area. Breeding progress has lagged for potato in comparison to other species due to its outbred, tetraploid genetics and a large number of complex traits. Wild potato species hold a potential key to exploit and utilize genetic resistance to pests and diseases, as well as tolerance to freezing, heat and drought. Implementation of GS as a tool for breeding potatoes has demonstrated a higher prediction accuracy and significant genetic gain. It allows genomic prediction and selection of traits at an early clonal generation stage in the absence of phenotyping hence reducing resource utilization. Development of GS-based approaches would radically enhance potato breeding by enabling early-stage selection for complex traits, thereby facilitating faster identification of elite breeding parents with desirable traits which would enhance potato breeding and development of new improved varieties.

Abbreviations

CIP	The International Potato Centre
GS	Genomic Selection
MAS	Marker Assisted Selection
GBS	Genotyping by Sequencing
NGS	Next Generation Sequencing
QTL	Quantitative Trait Loci
SNP	Single Nucleotide Polymorphism
BLUP	Best Linear Unbiased Prediction
GBLUP	Genomic best linear unbiased prediction
RKHS	Reproducing Kernel Hilbert Space
rrBLUP	Ridge regression best linear unbiased predictor
BGLR	Bayesian Generalized Linear Regression
BL	Bayesian LASSO
MAF	Minor Allele Frequency
GRM	Genomic Relationship Matrices
GEBVs	Genomic Estimated Breeding Values
GP	Genomic Prediction
GWAS	Genome-Wide Association Study
BAC	Bacterial Artificial Chromosome

CGIAR	Consultative Group for International Agricultural Research
DM	Doubled Monoploid
NPK	Nitrogen, phosphorus and potassium
EBN	Endosperm Balance Number
TPS	True Potato Seeds
LD	Linkage Decay
RFLP	Restriction fragment length polymorphism
GM	Genetically Modified
GMO	Genetically Modified Organism
TP	Training Populations

References

Arruda, M. P., Brown, P. J., Lipka, A. E., Krill, A. M., Thurber, C. et al. 2015. Genomic selection for predicting Fusarium head blight resistance in a wheat breeding program. *The Plant Genome* 8(3): 1–12.

Amadeu, R. R., Ferrão, L. F. V., Oliveira, I. D. B., Benevenuto, J., Endelman, J. B. and Munoz, P. R. 2020. Impact of dominance effects on autotetraploid genomic prediction. *Crop Science* 60(2): 656–665.

Ashraf, B. H., Jensen, J., Asp, T. and Janss, L. L. 2014. Association studies using family pools of outcrossing crops based on allele-frequency estimates from DNA sequencing. *Theoretical and Applied Genetics* 127(6): 1331–1341.

Bachem, C. W., van Eck, H. J. and de Vries, M. E. 2019. Understanding genetic load in potato for hybrid diploid breeding. *Molecular Plant* 12(7): 896–898.

Barone, A., Ritter, E., Schachtschabel, U., Debener, T., Salamini, F. et al. 1990. Localization by restriction fragment length polymorphism mapping in potato of a major dominant gene conferring resistance to the potato cyst nematode *Globodera rostocbiensis*. *Molecular and General Genetics MGG* 224(2): 177–182.

Barrell, P. J., Meiyalaghan, S., Jacobs, J. M. and Conner, A. J. 2013. Applications of biotechnology and genomics in potato improvement. *Plant Biotechnology Journal* 11(8): 907–920.

Bentley, A. R., Scutari, M., Gosman, N., Faure, S., Bedford, F. et al. 2014. Applying association mapping and genomic selection to the dissection of key traits in elite European wheat. *Theoretical and Applied Genetics* 127(12): 2619–2633.

Bernardo, R. and Yu, J. 2007. Prospects for genomewide selection for quantitative traits in maize. *Crop Science* 47(3): 1082–1090.

Bethke, P. C., Halterman, D. A. and Jansky, S. 2017. Are we getting better at using wild potato species in light of new tools? *Crop Science* 57(3): 1241–1258.

Bhat, J. A., Ali, S., Salgotra, R. K., Mir, Z. A., Dutta, S. et al. 2016. Genomic selection in the era of next generation sequencing for complex traits in plant breeding. *Frontiers in Genetics* 7: 221.

Bo, Z. H. U., Zhang, J. J., Hong, N. I. U., Long, G. U. A. N., Peng, G. U. O. et al. 2017. Effects of marker density and minor allele frequency on genomic prediction for growth traits in Chinese Simmental beef cattle. *Journal of Integrative Agriculture* 16(4): 911–920.

Bonierbale, M. W., Plaisted, R. L. and Tanksley, S. D. 1988. RFLP maps based on a common set of clones reveal modes of chromosomal evolution in potato and tomato. *Genetics* 120(4): 1095–1103.

Brown, C. R., Yang, C. P., Mojtahedi, H. A. S. S. A. N., Santo, G. S., Masuelli, R. I. C. A. R. D. O. et al. 1996. RFLP analysis of resistance to Columbia root-knot nematode derived from *Solanum bulbocastanum* in a BC 2 population. *Theoretical and Applied Genetics* 92(5): 572–576.

Byrne, S., Meade, F., Mesiti, F., Griffin, D., Kennedy, C. et al. 2020. Genome-wide association and genomic prediction for fry color in potato. *Agronomy* 10(1): 90.

Chen, Q., Li, H. Y., Shi, Y. Z., Beasley, D., Bizimungu, B., Goettel, M. S. et al. 2008. Development of an effective protoplast fusion system for production of new potatoes with disease and insect resistance using Mexican wild potato species as gene pools. *Canadian Journal of Plant Science* 88(4): 611–619.

CIP. 2019. CIP Annual Report 2018. Towards food system transformation. Lima, Peru. International Potato Center. 36 p. https://hdl.handle.net/10568/103463.

Clegg, M. T. 1990. Molecular diversity in plant populations. *Plant Population Genetics, Breeding, and Genetic Resources* 98–115

Combs, E. and Bernardo, R. 2013. Accuracy of genome-wide selection for different traits with constant population size, heritability, and number of markers. *The Plant Genome* 6(1): pp.plantgenome2012-11.

Darrier, B., Russell, J., Milner, S. G., Hedley, P. E., Shaw, P. D. et al. 2019. A comparison of mainstream genotyping platforms for the evaluation and use of barley genetic resources. *Frontiers in Plant Science* 10: 544.

de Bem Oliveira, I., Resende Jr, M. F., Ferrão, L. F. V., Amadeu, R. R., Endelman, J. B. et al. 2019. Genomic prediction of autotetraploids; influence of relationship matrices, allele dosage, and continuous genotyping calls in phenotype prediction. *G3: Genes, Genomes, Genetics* 9(4): 1189–1198.

Dufresne, F., Stift, M., Vergilino, R. and Mable, B. K. 2014. Recent progress and challenges in population genetics of polyploid organisms: an overview of current state-of-the-art molecular and statistical tools. *Molecular Ecology* 23(1): 40–69.

Eding, H. and Meuwissen, T. H. E. 2001. Marker-based estimates of between and within population kinships for the conservation of genetic diversity. *Journal of Animal Breeding and Genetics* 118(3): 141–159.

Edwards, M. D., Stuber, C. W. and Wendel, J. F. 1987. Molecular-marker-facilitated investigations of quantitative-trait loci in maize. I. Numbers, genomic distribution and types of gene action. *Genetics* 116(1): 113–125.

Enciso-Rodriguez, F., Douches, D., Lopez-Cruz, M., Coombs, J., de Los Campos, G. et al. 2018. Genomic selection for late blight and common scab resistance in tetraploid potato (*Solanum tuberosum*). *G3: Genes, Genomes, Genetics* 8(7): 2471–2481.

Endelman, J. B., Carley, C. A. S., Bethke, P. C., Coombs, J. J., Clough, M. E. et al. 2018. Genetic variance partitioning and genome-wide prediction with allele dosage information in autotetraploid potato. *Genetics* 209(1): 77–87.

FAO. 2010. The second Report on the State of the World's Plant Genetic Resources for Food and Agriculture. http://www.fao.org/3/i1500e/i1500e.pdf.

FAO. 2018. FAO statistical databases FAOSTAT http://www.fao.org/faostat/en/#home.

Fernando, R. L., Habier, D., Stricker, C., Dekkers, J. C. M., Totir, L. R. et al. 2007. Genomic selection. *Acta Agriculturae Scand Section A* 57(4): 192–195.

Fischer, M., Schreiber, L., Colby, T., Kuckenberg, M., Tacke, E. et al. 2013. Novel candidate genes influencing natural variation in potato tuber cold sweetening identified by comparative proteomics and association mapping. *BMC Plant Biology* 13(1): 1–15.

Garcia, A. A., Mollinari, M., Marconi, T. G., Serang, O. R., Silva, R. R. et al. 2013. SNP genotyping allows an in-depth characterisation of the genome of sugarcane and other complex autopolyploids. *Scientific Reports* 3(1): 1–10.

Gavrilenko, T., Antonova, O., Ovchinnikova, A., Novikova, L., Krylova, E. et al. 2010. A microsatellite and morphological assessment of the Russian National cultivated potato collection. *Genetic Resources and Crop Evolution* 57(8): 1151–1164.
Gebhardt, C., Mugniery, D., Ritter, E., Salamini, F., Bonnel, E. et al. 1993. Identification of RFLP markers closely linked to the H1 gene conferring resistance to *Globodera rostochiensis* in potato. *Theoretical and Applied Genetics* 85(5): 541–544.
Gebhardt, C., Ballvora, A., Walkemeier, B., Oberhagemann, P. and Schüler, K. 2004. Assessing genetic potential in germplasm collections of crop plants by marker-trait association: a case study for potatoes with quantitative variation of resistance to late blight and maturity type. *Molecular Breeding* 13(1): 93–102.
Gerard, D., Ferrão, L. F. V., Garcia, A. A. F. and Stephens, M. 2018. Genotyping polyploids from messy sequencing data. *Genetics* 210(3): 789–807.
Gidskehaug, L., Kent, M., Hayes, B. J. and Lien, S. 2011. Genotype calling and mapping of multisite variants using an Atlantic salmon iSelect SNP array. *Bioinformatics* 27(3): 303–310.
Godfray, H. C. J., Beddington, J. R., Crute, I. R., Haddad, L., Lawrence, D. et al. 2010. Food security: the challenge of feeding 9 billion people. *Science* 327(5967): 812–818.
Guo, X., Cericola, F., Fè, D., Pedersen, M. G., Lenk, I., Jensen, C. S. et al. 2018. Genomic prediction in tetraploid ryegrass using allele frequencies based on genotyping by sequencing. *Frontiers in Plant Science* 9: 1165.
Gupta, P. K., Kulwal, P. L. and Jaiswal, V. 2014. Association mapping in crop plants: opportunities and challenges. *Advances in Genetics* 85: 109–147.
Habier, D., Fernando, R. L. and Garrick, D. J. 2013. Genomic BLUP decoded: a look into the black box of genomic prediction. *Genetics* 194(3): 597–607.
Habyarimana, E., Parisi, B. and Mandolino, G. 2017. Genomic prediction for yields, processing and nutritional quality traits in cultivated potato (*Solanum tuberosum* L.). *Plant Breeding* 136(2): 245–252.
Hackett, C. A. and Luo, Z. W. 2003. Tetraploid Map: construction of a linkage map in autotetraploid species. *Journal of Heredity* 94(4): 358–359.
Hamilton, J. P., Hansey, C. N., Whitty, B. R., Stoffel, K., Massa, A. N. et al. 2011. Single nucleotide polymorphism discovery in elite North American potato germplasm. *BMC Genomics* 12(1): 1–12.
Hardigan, M. A., Laimbeer, F. P. E., Newton, L., Crisovan, E., Hamilton, J. P. et al. 2017. Genome diversity of tuber-bearing *Solanum* uncovers complex evolutionary history and targets of domestication in the cultivated potato. *Proceedings of the National Academy of Sciences* 114(46): E9999–E10008.
Hawkes, J. G. 1956. A revision of the tuber-bearing Solanums. *Scottish Plant Breeding Station Annual Report* 1956: 37–109.
Hawkes, J. G. 1963. A revision of the tuber-bearing Solanums II. *Scottish Plant Breeding Station Record* 1963: 76–181.
Hawkes, J. G. 1990. *The Potato: Evolution, Biodiversity and Genetic Resources*. Belhaven Press, London, UK.
Heffner, E. L., Jannink, J. L. and Sorrells, M. E. 2011. Genomic selection accuracy using multifamily prediction models in a wheat breeding program. *The Plant Genome* 4(1): 65–75.
Hehl, R., Faurie, E., Hesselbach, J., Salamini, F., Whitham, S. et al. 1999. TMV resistance gene N homologues are linked to *Synchytrium endobioticum* resistance in potato. *Theoretical and Applied Genetics* 98(3): 379–386.

Hein, I., McLean, K., Chalhoub, B. and Bryan, G. J. 2007. Generation and screening of a BAC library from a diploid potato clone to unravel durable late blight resistance on linkage group IV. *International Journal of Plant Genomics, 2007.*

Huang, S., Van Der Vossen, E. A., Kuang, H., Vleeshouwers, V. G., Zhang, N. et al. 2005. Comparative genomics enabled the isolation of the R3a late blight resistance gene in potato. *The Plant Journal* 42(2): 251–261.

International Food Policy Research Institute (IFPRI). 2019. IMPACT Projections of Food Production, Consumption, and Hunger to 2050, With and Without Climate Change: Extended Country-level Results for 2019 GFPR Annex Table 5. Washington, DC: IFPRI.

Jacobs, J. M., van Eck, H. J., Horsman, K., Arens, P. F., Verkerk-Bakker, B. et al. 1996. Mapping of resistance to the potato cyst nematode *Globodera rostochiensis* from the wild potato species *Solanum vernei. Molecular Breeding* 2(1): 51–60.

Jannink, J. L. 2007. Identifying quantitative trait locus by genetic background interactions in association studies. *Genetics* 176(1): 553–561.

Jansky, S. 2009. Breeding, genetics, and cultivar development. pp. 27–62. *In*: *Advances in Potato Chemistry and Technology*. Academic Press.

Jansky, S., Douches, D. and Haynes, K. 2018. Germplasm release: Three tetraploid potato clones with resistance to common scab. *American Journal of Potato Research* 95(2): 178–182.

Juyo Rojas, D. K., Soto Sedano, J. C., Ballvora, A., Léon, J., Mosquera Vásquez, T. et al. 2019. Novel organ-specific genetic factors for quantitative resistance to late blight in potato. *PloS One* 14(7): e0213818.

K Srivastava, R., Bollam, S., Pujarula, V., Pusuluri, M., Singh, R. B., Potupureddi, G. and Gupta, R. 2020. Exploitation of heterosis in pearl millet: a review. *Plants* 9(7): 807.

Kuhl, J. O. S. E. P. H., Hanneman, R. and Havey, M. 2001. Characterization and mapping of Rpi1, a late-blight resistance locus from diploid (1EBN) Mexican *Solanum pinnatisectum. Molecular Genetics and Genomics* 265(6): 977–985.

Kumari, M., Kumar, M. and Solankey, S. S. 2018. Breeding potato for quality improvement. Book: *Potato-From Incas to All Over the World* 37–59.

Kyriakidou, M., Achakkagari, S. R., López, J. H. G., Zhu, X., Tang, C. Y., Tai, H. H., ... Strömvik, M. V. 2020. Structural genome analysis in cultivated potato taxa. *Theoretical and Applied Genetics* 133(3): 951–966.

Li, L., Paulo, M. J., Strahwald, J., Lübeck, J., Hofferbert, H. R. et al. 2008. Natural DNA variation at candidate loci is associated with potato chip color, tuber starch content, yield and starch yield. *Theoretical and Applied Genetics* 116(8): 1167–1181.

Li, Y., Colleoni, C., Zhang, J., Liang, Q., Hu, Y. et al. 2018. Genomic analyses yield markers for identifying agronomically important genes in potato. *Molecular Plant* 11(3): 473–484.

Lindqvist-Kreuze, H., Gastelo, M., Perez, W., Forbes, G. A., de Koeyer, D. et al. 2014. Phenotypic stability and genome-wide association study of late blight resistance in potato genotypes adapted to the tropical highlands. *Phytopathology* 104(6): 624–633.

Liu, X., Wang, H., Wang, H., Guo, Z., Xu, X. et al. 2018. Factors affecting genomic selection revealed by empirical evidence in maize. *The Crop Journal* 6(4): 341–352.

Lorenz, A. J., Chao, S., Asoro, F. G., Heffner, E. L., Hayashi, T. et al. 2011. Genomic selection in plant breeding: knowledge and prospects. pp. 77–123. *In*: *Advances in Agronomy* (Vol. 110). Academic Press.

Lorenzana, R. E. and Bernardo, R. 2009. Accuracy of genotypic value predictions for marker-based selection in biparental plant populations. *Theoretical and Applied Genetics* 120(1): 151–161.

Marfil, C. F., Hidalgo, V. and Masuelli, R. W. 2015. *In situ* conservation of wild potato germplasm in Argentina: Example and possibilities. *Global Ecology and Conservation* 3: 461–476.

Meuwissen, T. H. E., Hayes, B. J. and Goddard, M. E. 2001. Prediction of total genetic value using genome-wide dense marker maps. *Genetics* 157: 1819–1829.
Meyer, R. C., Milbourne, D., Hackett, C. A., Bradshaw, J. E., McNichol, J. W. et al. 1998. Linkage analysis in tetraploid potato and association of markers with quantitative resistance to late blight (*Phytophthora infestans*). *Molecular and General Genetics MGG* 259(2): 150–160.
Michel, V., Julio, E., Candresse, T., Cotucheau, J., Decorps, C., Volpatti, R., Moury, B., Glais, L., Jacquot, E., de Borne, F. D. and Decroocq, V. 2019. A complex eIF4E locus impacts the durability of va resistance to Potato virus Y in tobacco. *Molecular Plant Pathology* 20(8): 1051–1066.
Moloney, C., Griffin, D., Jones, P. W., Bryan, G. J., McLean, K. et al. 2010. Development of diagnostic markers for use in breeding potatoes resistant to *Globodera pallida* pathotype Pa2/3 using germplasm derived from *Solanum tuberosum* ssp. andigena CPC 2802. *Theoretical and Applied Genetics* 120(3): 679–689.
Mosquera, T., Alvarez, M. F., Jiménez-Gómez, J. M., Muktar, M. S., Paulo, M. J. et al. 2016. Targeted and untargeted approaches unravel novel candidate genes and diagnostic SNPs for quantitative resistance of the potato (*Solanum tuberosum* L.) to *Phytophthora infestans* causing the late blight disease. *PLoS One* 11(6): e0156254.
Muir, W. M. 2007. Comparison of genomic and traditional BLUP-estimated breeding value accuracy and selection response under alternative trait and genomic parameters. *Journal of Animal Breeding and Genetics* 124(6): 342–355.
Naess, S. K., Bradeen, J. M., Wielgus, S. M., Haberlach, G. T., McGrath, J. M. and Helgeson, J. P. 2000. Resistance to late blight in *Solanum bulbocastanum* is mapped to chromosome 8. *Theoretical and Applied Genetics* 101(5): 697–704.
Neumann, A., Torstensson, G. and Aronsson, H. 2012. Nitrogen and phosphorus leaching losses from potatoes with different harvest times and following crops. *Field Crops Research* 133: 130–138.
Onogi, A., Watanabe, M., Mochizuki, T., Hayashi, T., Nakagawa, H. et al. 2016. Toward integration of genomic selection with crop modelling: the development of an integrated approach to predicting rice heading dates. *Theoretical and Applied Genetics* 129(4): 805–817.
Ortiz, R. 2020. *Genomic-Led Potato Breeding for Increasing Genetic Gains: Achievements and Outlook.*
Osborn, T. C., Pires, J. C., Birchler, J. A., Auger, D. L., Chen, Z. J. et al. 2003. Understanding mechanisms of novel gene expression in polyploids. *Trends in Genetics* 19(3): 141–147.
Pajerowska-Mukhtar, K., Stich, B., Achenbach, U., Ballvora, A., Lübeck, J. et al. 2009. Single nucleotide polymorphisms in the Allene Oxide Synthase 2 gene are associated with field resistance to late blight in populations of tetraploid potato cultivars. *Genetics* 181(3): 1115–1127.
Parisseaux, B. and Bernardo, R. 2004. *In silico* mapping of quantitative trait loci in maize. *Theoretical and Applied Genetics* 109(3): 508–514.
Paterson, A. H., Lander, E. S., Hewitt, J. D., Peterson, S., Lincoln, S. E. et al. 1988. Resolution of quantitative traits into Mendelian factors by using a complete linkage map of restriction fragment length polymorphisms. *Nature* 335(6192): 721–726.
Pham, G. M., Newton, L., Wiegert-Rininger, K., Vaillancourt, B., Douches, D. S. et al. 2017. Extensive genome heterogeneity leads to preferential allele expression and copy number-dependent expression in cultivated potato. *The Plant Journal* 92(4): 624–637.
Pham, G. M., Hamilton, J. P., Wood, J. C., Burke, J. T., Zhao, H. et al. 2020. Construction of a chromosome-scale long-read reference genome assembly for potato. *Gigascience* 9(9): giaa100.
Piepho, H. P. 2009. Ridge regression and extensions for genomewide selection in maize. *Crop Science* 49(4): 1165–1176.

Pimentel, E. C. G., Edel, C., Emmerling, R. and Götz, K. U. 2015. How imputation errors bias genomic predictions. *Journal of Dairy Science* 98(6): 4131–4138.

Potato Genome Sequencing Consortium. 2011. Genome sequence and analysis of the tuber crop potato. *Nature* 475(7355): 189.

Premanandh, J. 2011. Factors affecting food security and contribution of modern technologies in food sustainability. *Journal of the Science of Food and Agriculture* 91(15): 2707–2714.

Priegnitz, U., Lommen, W. J., van der Vlugt, R. A. and Struik, P. C. 2019. Impact of positive selection on incidence of different viruses during multiple generations of potato seed tubers in Uganda. *Potato Research* 62(1): 1–30.

Ritter, E., Debener, T., Barone, A., Salamini, F., Gebhardt, C. et al. 1991. RFLP mapping on potato chromosomes of two genes controlling extreme resistance to potato virus X (PVX). *Molecular and General Genetics MGG* 227(1): 81–85.

Simko, I., Haynes, K. G., Ewing, E. E., Costanzo, S., Christ, B. J. et al. 2004. Mapping genes for resistance to *Verticillium alboatrum* in tetraploid and diploid potato populations using haplotype association tests and genetic linkage analysis. *Molecular Genetics and Genomics* 271(5): 522–531.

Slater, A. T., Cogan, N. O. I., Hayes, B. J., Schultz, L., Dale, M. F. B. et al. 2014a. Improving breeding efficiency in potato using molecular and quantitative genetics. *Theoretical and Applied Genetics* 127: 2279–2292. doi:10.1007/ s00122-014-2386-8.

Slater, A. T., Wilson, G. M., Cogan, N. O. I., Forster, J. W., Hayes, B. J. et al. 2014b. Improving the analysis of low heritability complex traits for enhanced genetic gain in potato. *Theoretical and Applied Genetics* 127: 809–820. doi:10.1007/ s00122-013-2258-7.

Slater, A. T., Cogan, N. O., Forster, J. W., Hayes, B. J., Daetwyler, H. D. et al. 2016. Improving genetic gain with genomic selection in autotetraploid potato. *The Plant Genome* 9(3): 1–15.

Song, J., Bradeen, J. M., Naess, S. K., Helgeson, J. P., Jiang, J. et al. 2003. BIBAC and TAC clones containing potato genomic DNA fragments larger than 100 kb are not stable in Agrobacterium. *Theoretical and Applied Genetics* 107(5): 958–964.

Spooner, D. M., Núñez, J., Trujillo, G., del Rosario Herrera, M., Guzmán, F. et al. 2007. Extensive simple sequence repeat genotyping of potato landraces supports a major reevaluation of their gene pool structure and classification. *Proceedings of the National Academy of Sciences* 104(49): 19398–19403.

Stich, B. and Van Inghelandt, D. 2018. Prospects and potential uses of genomic prediction of key performance traits in tetraploid potato. *Frontiers in Plant Science* 9: 159.

Sverrisdóttir, E., Byrne, S., Sundmark, E. H. R., Johnsen, H. Ø., Kirk, H. G. et al. 2017. Genomic prediction of starch content and chipping quality in tetraploid potato using genotyping-by-sequencing. *Theoretical and Applied Genetics* 130(10): 2091–2108.

Sverrisdóttir, E., Sundmark, E. H. R., Johnsen, H. Ø., Kirk, H. G., Asp, T. et al. 2018. The value of expanding the training population to improve genomic selection models in tetraploid potato. *Frontiers in Plant Science* 9: 1118.

Uitdewilligen, J. G., Wolters, A. M. A., Bjorn, B., Borm, T. J., Visser, R. G. et al. 2013. A next-generation sequencing method for genotyping-by-sequencing of highly heterozygous autotetraploid potato. *PloS One* 8(5): e62355.

United Nations, Department of Economic and Social Affairs, Population Division. 2019. World Population Prospects 2019: Ten Key Findings. https://population.un.org/wpp/Publications/ accessed 2020.

United Nations, Department of Economic and Social Affairs, Population Division. 2019. World Population Prospects 2019: Highlights. ST/ESA/SER.A/423.

Van der Voort, J. R., Wolters, P., Folkertsma, R., Hutten, R., Van Zandvoort, P. et al. 1997. Mapping of the cyst nematode resistance locus Gpa2 in potato using a strategy based on comigrating AFLP markers. *Theoretical and Applied Genetics* 95(5): 874–880.

Van Der Vossen, E. A., Van Der Voort, J. N. R., Kanyuka, K., Bendahmane, A., Sandbrink, H. et al. 2000. Homologues of a single resistance-gene cluster in potato confer resistance to distinct pathogens: a virus and a nematode. *The Plant Journal* 23(5): 567–576.

VanRaden, P. M. 2008. Efficient methods to compute genomic predictions. *Journal of Dairy Science* 91(11): 4414–4423.

VanRaden, P. M., O'Connell, J. R., Wiggans, G. R. and Weigel, K. A. 2011. Genomic evaluations with many more genotypes. *Genetics Selection Evolution* 43(1): 10.

Von Braun, J. 2007. The world food situation: new driving forces and required actions. *Intl. Food Policy Res. Inst. Washington DC.*

Vos, P. G., Uitdewilligen, J. G., Voorrips, R. E., Visser, R. G., van Eck, H. J. et al. 2015. Development and analysis of a 20K SNP array for potato (*Solanum tuberosum*): an insight into the breeding history. *Theoretical and Applied Genetics* 128(12): 2387–2401.

Walpers, W. G. 1884. *Repertorium botanices sytematicae*. Vol 3. Hofmeister, Leipzig, Germany.

Wang, X., Su, G., Hao, D., Lund, M. S., Kadarmideen, H. N. et al. 2020. Comparisons of improved genomic predictions generated by different imputation methods for genotyping by sequencing data in livestock populations. *Journal of Animal Science and Biotechnology* 11(1): 1–12.

Wilson, S., Zheng, C., Maliepaard, C., Mulder, H. A., Visser, R. G. et al. 2021. Understanding the effectiveness of genomic prediction in tetraploid potato. *Frontiers in Plant Science* 1634.

Zhang, A., Wang, H., Beyene, Y., Semagn, K., Liu, Y. et al. 2017. Effect of trait heritability, training population size and marker density on genomic prediction accuracy estimation in 22 bi-parental tropical maize populations. *Frontiers in Plant Science* 8: 1916.

Zhang, C., Wang, P., Tang, D., Yang, Z., Lu, F. et al. 2019. The genetic basis of inbreeding depression in potato. *Nature Genetics* 51(3): 374–378.

10

Genomic Selection Schemes for Long-term Gain

Ani A Elias[1,*] and *Jean-Luc Jannink*[2,3]

ABSTRACT

Genetic gain from genomic selection is equally important for its predictive ability when it comes to a crop improvement program. Genetic gain is quantified as the change in mean phenotypic value of a trait between parent and offspring generations. The gain depends on the genetic variance available in the breeding population, selection intensity, and time required to complete a selection in addition to the prediction accuracy. There are many modifications in conventional genomic selection methods to increase the genetic variance in the population retaining the favorable alleles for more than one generation. In this chapter, we introduce most of these methods such as optimum haploid value, optimum population value, predicted cross value, genomic mating and look-ahead mate selection. Validating the adaptability and heritability of newly developed genotypes in their target environments is an inevitable process in crop improvement programs. Therefore, we extend this chapter to introduce methods for designing and analyzing multi-environmental trials facilitating optimal use of resources while performing genomic selection.

[1] Institute of Forest Genetics and Tree Breeding, Coimbatore, Tamil Nadu, India - 641002.

[2] Department of Plant Breeding and Genetics, School of Integrative Plant Sciences, Cornell University, Ithaca, New York, USA - 14850.

[3] USDA-ARS, Ithaca, New York, USA - 14850.

* Corresponding author: anianna01@gmail.com

Background

The success of genomic selection (GS) and prediction (GP) is measured by its accuracy, i.e., the correlation between the observed and estimated value of the trait of interest. Many factors such as the linkage between markers and genes of interest, population size, genomic relationship among individuals, and heritability influence the accuracy (Zhang et al., 2019). Prediction accuracy also depends on the statistical model (Lee et al., 2017), minor allele frequency (MAF) as low MAF markers reduce the prediction accuracy in BayesR (Zhang et al., 2019), whether the crop is inbred or outbred (Howard et al., 2014), and having a pronounced population structure or not (Isidro et al., 2015).

From a breeder's perspective, genetic gain is also an important scale in measuring the long-term success of GS. Genetic gain can be defined as the amount of increase in (mean) performance achieved through artificial selection. Genetic gain is measured by the difference between the mean phenotypic value of a trait in the parent and offspring generations (Falconer and Mackay, 2009). In GS, genetic gain, ΔG, can be measured as follows:

$$\Delta G = i\ r\ \sigma_A/t$$

where i is the selection intensity which is the mean deviation of selection in the phenotypic observation; r is the selection accuracy which is the correlation between the true and estimated breeding values (BV); σ_A is the additive genetic standard deviation; t is the breeding cycle time.

Factors affecting genetic gain are (i) genetic variation available in the breeding population as increased variation increases the opportunity for recombination and a higher gain, (ii) heritability of the trait, (iii) selection pressure considering the diversity and inbreeding, and (iv) time required to complete the breeding pipeline (Daetwyler et al., 2015; Jannink et al., 2010). Favorable genetic gain can be achieved from GS through selection of genetically favorable as well as diverse parents for the next generation. Multiple genome based reference, a pangenome, can also be used for unlocking genetic variation, rare alleles, favorable alleles, and haplotyes hidden in diverse germplasm collections (Xu, 2012; Golicz et al., 2016). Heritability depends on the additive genetic variance and its estimation can be refined using statistical models accounting for non-additive genetic variances as well as field variations. Selection intensity can be improved using modifications in the genotyping and phenotyping. Breeding cycle can be reduced using GS and modifications in the statistical model and training population design.

Genetic variation is the foundation of genetic gain. The gain in the selection response from individual genes is proportional to the genetic variance explained by them. Therefore, for long term genetic gain one may need to know the details of the genes causing the genetic variation (Goddard,

2009). When it comes to complex traits, however, many genes with small genetic effects explain the genetic variance and therefore each of the genes contribute a little to the genetic gain. As a workaround, high density markers throughout the genome are used in GS when information on causative genes is not available. This strategy assumes that QTL can be located anywhere in the genome and on using dense markers all the QTL are in complete LD with at least one marker (Daetwyler et al., 2015).

Use of bi-parental populations is popular in plant breeding schemes. Bi-parental populations have limitations in allelic diversity which can result in a faster decline of genetic variance over generations. Increase in genetic gain for the short-term can be achieved through GS by increasing the selection intensity based on individual BV. The increased selection intensity without considering the population genetic diversity, however, can reduce the genetic variance, increase inbreeding and co-ancestry, and can also result in the loss of low-frequency variants as the generations proceed (Jannink et al., 2010; Akdemir and Sánchez, 2016).

Including multiple bi-parental populations and use of multi-parent population (Verhoeven et al., 2006) are suggested as work-around solutions for limited allelic diversity. But across different bi-parental populations with different allelic diversity, the prediction accuracy will be low. Jannink et al. (2010) suggested to have a different GS model for each bi-parental population in such a scenario. In a simulation study done by de Roos et al. (2009), increased prediction accuracy was achieved when multiple genomically diverse populations were combined in one training population at the cost of increased marker density. From the practical point of view, multiple parent populations may not be an attractive option in plant breeding as the mean phenotypic value of multiple parents may be lower than that of bi-parents. The advantage of a multi-parent breeding scheme is the added allelic diversity and thus increased genetic diversity.

Many modified GS methods are proposed to achieve long-term genetic gain through increased genetic diversity. This chapter is intended to provide brief explanations on such modified GS methods. For simplicity, the base GS model is referred to as the conventional GS (CGS) model from now onwards in this chapter.

Modified GS methods

Optimum haploid value (OHV)

The optimum haploid value (OHV) (Daetwyler et al., 2015) was proposed as an alternate method to CGS for evaluating the potential of a genotype to be a parent for the next generation by estimating the BV. OHV relies on the recombination process during meiosis resulting in haplotypes with favorable alleles. A haplotype can be defined as a set of genes that is inherited in a chromosome within an organism from a single parent or ancestral line (Yunbi Xu et al., 2017). OHV facilitates the selection of the best double haploid (DH). Similar to a genome estimated BV (GEBV), OHV measures the individual BV but on the DHs. The following summarization is based on the conceptualization and evaluation done by Daetwyler et al. (2015).

The ploidy level determines the number of haploid genomes an organism carries. It is easy to demonstrate the OHV in a diploid which can be extended to polyploid organisms. In OHV selection, haplotype values (HV) are calculated for all the haplotypes in a genomic segment,

$$HV = \sum_{k=1}^{m} h_k \beta_k$$

where m is the total number of loci in the haplotype, k is the locus within the segment, and h_k is the individual's haplotype at locus k. The length of the haplotype considered can be varied and optimized to maximize the genetic gain. To calculate the OHV, the best HV in each segment is summed over segments

$$OHV = 2\sum_{o=1}^{n} max(HV_O)$$

where n is the number of OHV segments, and o is the genomic segment. Multiplication by 2 allows a direct comparison with GEBV and fully shows the genetic level of the potential DH.

OHV of an individual is always equal to or greater than the GEBV. While considering the Mendelian segregation and similarity in marker effect estimation, the difference between GEBV and OHV is that GEBV is the realized breeding value of an individual while OHV is a futuristic breeding value of an individual when the best chromosome segments are inherited and doubled at each position in the genome. For an inbred, the OHV is equal to the GEBV because the best haploid that can be produced is constant regardless of which haplotype is chosen. Therefore, OHV is valuable in implementing in segregating individuals. As a procedure of OHV, first, outbred individuals are genotyped and their HV and OHV are calculated. Second, the best OHV plants are selected for DH production. Third, the double haploid seedlings are genotyped and their GEBVs are estimated to be used for the next generation or to be selected as the elite variety. Therefore, in OHV, genotyping is done in two generations; haplotyping is done to enhance the pipeline of the GP through the GEBV calculation. The OHV and GEBV calculation will be repeated in each generation where new recombination events are expected.

Selections based on OHV resulted in increased genetic gain over CGS in scenarios where sufficient recombination events were accumulated. The selection is focused on recombinations in smaller segments of the haplotype. Based on the simulation data, there was a clear increase in genetic gain of OHV as the number of offsprings increased which favored the chance of recombination and thus the increase in HV. The genetic gain for OHV is long term compared to that from CGS as OHV takes advantage of the recombination process. In the breeding population, however, the overall GEBV was observed to be lower compared to that in CGS due to the diversity. Since the selection is based on fewer segments of chromosomes, it is possible that the phenotype produced does not meet the expected threshold of a complex trait as a large number of genes (even beyond the targeted segments in OHV) have small contributions to it.

A significant increase in genetic diversity was also observed with OHV compared to the GEBV. In GEBV, individuals are selected based on the sum of all the allele effects whereas in OHV it is based on the haplotype value of each segment of the genome. So, in GEBV, presence of unfavorable alleles reduces the ranking of an individual and therefore GEBV favors the presence of a less diverse set of individuals after selection. In OHV, the ranking is based on the haploid value of a segment; if the haplotype value of a segment has favorable alleles, OHV takes only that into account and ignores the segment with lesser merit. The individual will be selected into the pipeline of breeding thus increasing the diversity in selection. The haplotype maintained may be inferior, because it carries several unfavorable alleles but eventually through recombination these haplotypes release favorable alleles into the breeding population and then their frequency can be increased by selection.

Optimum population value (OPV)

The concept of OPV was introduced by Goiffon et al. (2017) as a combination of genotype building (GB) (Kemper et al., 2012) and OHV and applied to a subset of individuals in a breeding population. GB is a group selection and mating method to find a minimal group based on segregating favorable chromosome segments. OPV adopts the idea of group selection from GB but the group is identified as individuals that maximize the selection limit based on haploid values. The improvement of OPV over OHV is that it selects individuals with the highest haploid value while OPV selects a group. Because of the individual metric-based selection, it is possible that OHV will discard individuals with rare favorable alleles that are masked by many unfavorable alleles (Daetwyler et al., 2015). OPV allows the inclusion of individuals that have low genetic merit but contain rare favorable alleles by selecting complimentary individuals in the group. Both OPV and OHV, however, maintain rare favorable alleles in the breeding population facilitating long term genetic gain. OPV outperforms OHV and GB and all the three methods outperform CGS as shown by the results of simulation studies (Goiffon et al., 2017). The summarization of concept of OPV follows.

OPV considers the haploid values of individuals but evaluates the merit of potential progeny of a subset of selection candidates as in GB selection. OPV can be considered as a method to determine the GEBV of the best possible progeny produced after many generations by a breeding population.

$$OPV(S) = M.\sum_{b=1}^{B} \sum_{l \in H(b)} A_{l,m,n} e_l$$

where S is the optimal breeding population, n is the number of individuals, M is the ploidy of the plants, B is the number of haplotype blocks per chromosome, H is the set of marker loci that belong to a haplotype block, e_l is the effect of having the major allele at locus l.

By changing the number of haplotype blocks, B, short-term or long-term gain is achieved. B determines the number of recombination events. Larger the number, higher the expected recombination events and longer the genetic gain. Also, by removing individuals with lower GEBV values harboring favorable alleles result in short-term genetic gain while including those individuals can increase the long-term genetic gain.

Selecting groups is better as it favors more genetic variation to be included and possible recombination events. And selecting based on the genetic segments is also an improvement when aiming for long-term genetic gain. There are computational challenges involved in selecting groups of individuals in this fashion such as solving a combinatorial optimization problem considering the number of potential recombination and mating events. However, the group selection method is a better strategy for longer term genetic gain by reducing inbreeding and increasing the genetic gain for a longer period. OPV selects the best group of individuals based on their interactive effect and calculates the GEBV of the best possible progeny from this group after an unlimited number of generations.

All three methods (OHV, OPV, and GB) consider the effects of recombination on the genetic merit of future individuals by focusing on the GEBV of segments of the genome (the haploid block) instead of the total GEBV. These methods can identify and select for segments that contain rare favorable alleles which can increase the genetic gain for longer periods of time. For example, complimentary donors can be identified in the context of donor × recipient crosses by accounting for the recombination frequencies and LD between markers at the haplotype segments (Allier et al., 2020).

Predicted cross value (PCV)

The predicted cross value (PCV) (Han et al., 2017) calculates the probability of producing an ideal genotype consisting of only favorable alleles for the next two generations from a donor × recipient cross based on the specific combining ability. In GEBV and OHV parents are selected based on the highest score criterion for an individual. In addition the OPV is about the general combining ability of a group. In PCV, the probability is calculated for a pair of individuals where desirable alleles are accumulated. The probability relies on the precise estimation of favorable alleles (Lehermeier et al., 2017).

A summarization of PCV based on the conceptualization of Han et al. (2017) follows. A vector of recombination factors between random pairs of individuals is used to calculate the probability of producing a gamete with no undesirable alleles thus giving an ideal genotype in the coming generation. Therefore, we can select those sets of parents with multiple desirable alleles with the highest likelihood of producing a gamete for the next two generations with the least number of undesirable alleles. Unlike in GEBV and OHV which are based on individual scores, PCV calculates the score matrix based on potential parent pairs by identifying complementary alleles. In PCV, two individuals are treated as a unique set and the likelihood value is estimated for its ability to produce highly desirable gametes. In other words, the GEBV and OHV both give indices for the general combining ability while PCV is about the specific combining ability. PCV saves time compared to traditional crossing methods such as back crossing in the introgression of desirable alleles. Simulations showed that PCV outperforms GEBV and OHV in long-term genetic gain.

Genomic mating (GM)

Genomic mating (GM) is an optimization method proposed by Akdemir and Sanchez (2016) as an alternative to CGS. The method uses the GEBV and coefficient of co-ancestry to optimize the mating pair. The idea of genetic mating, known as look ahead mate selection (LAMS) schemes have been used in animal breeding through simulations since the 1990s (Kinghorn and Shepherd, 1994; Hayes et al., 1998; Shepherd and Kinghorn, 1998; Hayes et al., 2009). These methods, however, used pedigree based co-ancestry matrices and additive genetic value. Based on these values, a mate selection index (MSI) and a mate selection algorithm were developed to identify the pair which can maximize the merit of progeny. The GM uses genomic data with high density marker information to estimate BV, Mendelian effect, and co-ancestry to incorporate into MSI.

GM focuses on the mating compatibility of individuals for long term genetic gain and uses genomically estimated within—cross variance to find complementary mating parents. Exploitation of genetic markers through crossing and selection is a combinatorial optimization problem in a genetic context. The number of possible crossings grow exponentially with the number of parental genotypes which makes the task of this optimization challenging (de Beukelaer et al., 2015).

Akdemir and Sanchez (2016) provided new solutions to MSI considering genetic gain, variance, and inbreeding. They also provide a genetic algorithm that can find good solutions to finding complementary mates for sustainable genetic gain for complex traits. Their solution uses additive genetic variance only. For the solution of the mating problem, the authors derived a measure called the risk of mating plan and combining it with the inbreeding measure, the mating problem solution is derived as under:

$$r(\lambda_1, \lambda_2, P_{c\times n}) = -Risk(\lambda_1, P_{c\times n}) + \lambda_2 \times Inbreeding(P_{c\times n})$$

where $\lambda_2 \geq 0$ is the parameter controlling the amount of co-ancestry in the progeny, and the minimization is over the space of the mating matrix P with each row having two half values at positions corresponding to two distinct parents or a value of 1 for selfed parent, c is the number of children from n parents; each row of P has two half values at positions corresponding to two distinct parents or only a value at the position corresponding to the selfed parent; λ_2 controls allele diversity and λ_1 controls allele heterozygosity, which is the risk parameter and it is calculated based on the marker matrix and the expected number of beneficial alleles in the children for each parent pair in the mating matrix. Inbreeding is calculated using the variance-covariance of the progeny based on P and the realized genomic relationship matrix of the parents. The solution makes assumptions such as (i) diploid behavior in meiosis, (ii) uncorrelated distribution of genes, (iii) absence of allelic interaction, (iv) absence of multiple alleles at targeted loci controlling the trait, and (v) absence of genotype by environment interaction.

The productivity of a crop is based on multiple characteristics such as yield, disease resistance with different breeding goals. A successful breeding scheme is based on the selection of individuals considering these different breeding goals. GM is one such method of simultaneous mate optimization and selection (Akdemir and Sánchez, 2016; Akdemir et al., 2019) which is applicable in single and multi-trait scenarios. Such an optimal mating scheme can be used in increasing genetic variance in the progenies which can decrease inbreeding and co-ancestry while simultaneously attaining genetic gain. This scheme can result in increased additive genetic variance in the progeny for a given set of parents when compared to a random mating approach. Based on the simulation study conducted by the authors, a clear advantage on genetic gain is attained by GM compared to phenotypic selection and CGS (Akdemir and Sánchez, 2016). The script for the GM optimization algorithm is available from the authors. Another mate allocation and selection optimization program, AlphaMate, with executable versions is available in http://www.AlphaGenes.roslin.ed.ac.uk/AlphaMate (Gorjanc and Hickey, 2018). Recently, Wolf et al.

(2021) demonstrated the use of the non-additive effect also for GM in cassava quantifying significant inbreeding depression. They implemented the core functions in an R package predCrossVar.

Genomic look-ahead mate selection

The GM proposed by Akdemir and Sanches (2016) is for finding the best mating set for optimal performance in the next generation. Genomic LAMS is a method where mating selection is optimized for two generations ahead (grand progeny). Genomic LAMS introduced in plant breeding by Moeinizade et al. (2019) outperformed CGS, OHV, and OPV by achieving more genetic gain and preserving more genetic diversity in the simulated breeding program. Later, the authors extended the single trait LAMS (ST-LAMS) to multi-trait LAMS (MT-LAMS) retaining the advantages of ST-LAMS (Moeinizade et al., 2020).

MT-LAMS maximizes a key trait, for example, yield, while keeping the other traits within bounded threshold values. The MT-LAMS algorithm focuses on the main trait when most of the individuals become acceptable for all the traits of interest. So, there is a dynamic adjustment in mating and selection decision until all the traits are within the boundaries of thresholds in every generation. Moeinizade et al. (2020) claim that a unique advantage of MT-LAMS is that it focusses on the genetic gain of the progeny in the terminal generation such as a grand-progeny and thus maximizes the expected GEBV of that progeny.

LAMS makes an optimal trade-off between short- and long-term genetic gain to achieve the highest genetic gain within a specified time. LAMS provides optimized solutions in (i) deciding a deadline for the breeding pipeline, (ii) selecting an explicit mating strategy for pairs as the mating scheme affects the genetic gain (Akdemir and Sánchez, 2016; Wang et al., 2018) and, (iii) by allocating a fixed total budget over a period of time to achieve the best and final outcome which is a strategic decision (Lorenz, 2013; Lin et al., 2017).

The optimization of selection and mating, resource allocation, and time management is a computationally challenging method due to complex factors such as recombination frequencies, mating strategies, time management and resource allocations. Moeinizade et al. (2019) introduced a simulation optimization algorithm in LAMS dealing with these challenges.

The LAMS method is formulated as follows:

$$\int^{LAS}(x, y, r, T - t)$$

$$x_n = \sum_{j=1}^{N} y_{n,j} n \in \{1,\ldots,N\}, \quad y_{i,j} \in \{0,1\}, \quad r \in \{0, 0.5\}^{L-1}$$

where x is the selected breeding parents; $y_{i,j}$ is a binary variable for the mating decision that shows whether individual i is mated with individual j ($y_{i,j} = 1$) or not ($y_{i,j} = 0$); r is the recombination frequency factor; T is the terminal generation and t is the current generation. The variables, x and y are input variables that need to be optimized by the model; r and $(T - t)$ are parameters that the model needs to consider when searching for the optimal solution. The model maximizes the expected GEBV of the best offspring in the terminal generation.

In CGS, OHV, and OPV unlimited time and resources are assumed while estimating genetic gain. LAMS is computationally less challenging compared to OHV and OPV through the optimization of time and resources. Time optimization is achieved by defining terminal generation and resource optimization by pre-deciding the mating pairs and thus limiting the number of recombination frequencies of haplotype blocks. The mating pairs are identified based on the higher predictive genetic variation and probability of producing outstanding progenies (Moeinizade et al., 2020).

Interaction models

Predicting the performance in one location using the data from another location is a crucial step in crop improvement programs. Multi-environmental trials (MET) for assessing genotype by environment ($G \times E$) has an important role in plant breeding for selecting high–performing and stable lines across environments. Burgueno et al. (2012) were the first to use marker and pedigree based GBLUP for assessing $G \times E$ for genomic predictions. Heslot et al. (2014) used mechanistic models incorporated in the statistical model for $G \times E$ assessment. Jarquin et al. (2014) developed a reaction norm model using a variance co-variance structure of marker and environmental covariates.

There are interaction models with SNP marker effects accounting for the environment specific variance in a marker. With positive correlation among the locations, marker effect × environmental ($M \times E$) models perform better than single location or multi-location $G \times E$ models. Also, with multi-trait analysis, the $M \times E$ model performs well when the traits are highly positively correlated (Lopez-Cruz et al., 2015; Crossa et al., 2016; Gapare et al., 2018). Pre-selection of markers for $M \times E$ can be done through association mapping (Vallejo et al., 2017). Markers can also be included using the differential shrinkage property of the model or using the variable selection method (de los Campos et al., 2013).

Sparse testing

Replicating all genotypes in all the environments in MET is impractical when a large number of genotypes are present and the resources are not adequate to accommodate all of them (Butler et al., 2014; Oakey et al., 2006). There are augmented designs and their statistical models where control genotypes are replicated in all environments and test genotypes are represented once or replicated in a few of the environments (Smith et al., 2006; Walter T. Federer, 1961; 2005; Federer and Raghavarao, 1975). All these models, however, do not explore the use of genomic information. The unobserved interaction effect can be predicted using genomic information in an interaction model. The prediction accuracy depends on the (i) number of overlapping genotypes, (ii) number of environments in which each genotype is grown, and (iii) the statistical model. Shared genomic information provides connectivity in augmented designs allowing MET to be conducted earlier in the breeding program (Endelman et al., 2014). The strategy can also help in maximizing the genetic gain in preliminary yield trials. Sparse testing (Jarquin et al., 2020) is another genomic information-based statistical method developed for augmented designs. In sparse testing, predictions on unobserved $G \times E$ are estimated based on the shared information among the genotypes and environments using the genomic relationship matrix and environmental correlation. The predictions can be made for genotypes not evaluated in some environments but others as well as for genotypes that are never tested in any environment.

Based on the results, Jarquin et al. (2020) concluded that the prediction accuracy increased with the increase in the number of common genotypes in all the environments and it decreased with a decrease in the sample size. Allocating common genotypes in MET increases the prediction accuracy while augmenting the design with partial replications and for non-tested genotypes. Based on the prediction improvement, authors recommended up to 25% of the total number of genotypes as common genotypes.

Traits with moderate to high heritability are effective in sparse testing compared to those with low heritability as demonstrated by Santantonio et al. (2020). In addition to heritability and number of common genotypes, field resource availability, amount of historical data available for training prediction models, structure of breeding populations, and access to cost-effective genotyping platforms are all critical deciding factors on choosing an optimal augmented design for early field testing in crop improvement programs (Santantonio et al., 2020) facilitating genetic gain from GS. The Genomic Open-source Breeding informatics initiative (GOBii) (www.gobiiproject.org) and Excellence in Breeding Platform (EiB) (www.excellenceinbreeding.org) are two multi-institutional undertakings which work with public sector breeding programs in merging and storing data, facilitating cost-effective genotyping, tracking samples, implementing advanced analysis pipelines, and guiding in time-bound decision making.

Conclusions

Modified versions of the GS model account for not only the accuracy in prediction but also the long-term genetic gain from the crop improvement program. This idea of long-term gain is achieved through identifying compatible mating pairs of parents to maintain the genetic diversity and to retain rare but favorable alleles over generations. As in Nassim Taleb's 'The Black Swan', 'the impact of the highly improbable' exists and the "the inability to predict outliers implies the inability to predict the course of history".

Acknowledgments

Ani A Elias is thankful to the Department of Biotechnology (DBT), Government of India for the Ramalingaswami fellowship to continue her research work in GS which helped her to write this chapter.

References

Akdemir, Deniz and Julio I. Sánchez. 2016. Efficient breeding by genomic mating. *Frontiers in Genetics* 7(November). https://doi.org/10.3389/fgene.2016.00210.

Akdemir, Deniz, William Beavis, Roberto Fritsche-Neto, Asheesh K. Singh et al. 2019. Multi-objective optimized genomic breeding strategies for sustainable food improvement. *Heredity* 122(5): 672–83. https://doi.org/10.1038/s41437-018-0147-1.

Allier, Antoine, Simon Teyssèdre, Christina Lehermeier, Alain Charcosset et al. 2020. Genomic prediction with a maize collaborative panel: identification of genetic resources to enrich elite breeding programs. *Theoretical and Applied Genetics* 133(1): 201–15. https://doi.org/10.1007/s00122-019-03451-9.

Beukelaer, Herman de, Geert de Meyer and Veerle Fack. 2015. Heuristic exploitation of genetic structure in marker-assisted gene pyramiding problems. *BMC Genetics* 16(1): 2. https://doi.org/10.1186/s12863-014-0154-z.

Burgueño, Juan, Gustavo de los Campos, Kent Weigel, José Crossa et al. 2012. Genomic prediction of breeding values when modeling genotype × environment interaction using pedigree and dense molecular markers. *Crop Science* 52(2): 707–19. https://doi.org/10.2135/cropsci2011.06.0299.

Butler, David G., Alison B. Smith, Brian R. Cullis et al. 2014. On the design of field experiments with correlated treatment effects. *Journal of Agricultural, Biological, and Environmental Statistics* 19(4): 539–55. https://doi.org/10.1007/s13253-014-0191-0.

Campos, Gustavo de los, John M. Hickey, Ricardo Pong-Wong, Hans D. Daetwyler et al. 2013. Whole-genome regression and prediction methods applied to plant and animal breeding. *Genetics* 193(2): 327–45. https://doi.org/10.1534/genetics.112.143313.

Crossa, José, Gustavo de los Campos, Marco Maccaferri, Roberto Tuberosa, J. Burgueño et al. 2016. Extending the marker × environment interaction model for genomic-enabled prediction and genome-wide association analysis in durum wheat. *Crop Science* 56(5): 2193–2209. https://doi.org/10.2135/cropsci2015.04.0260.

Daetwyler, Hans D., Matthew J. Hayden, German C. Spangenberg, Ben J. Hayes et al. 2015. Selection on optimal haploid value increases genetic gain and preserves more genetic diversity relative to genomic selection. *Genetics* 200(4): 1341–48. https://doi.org/10.1534/genetics.115.178038.

Endelman, Jeffrey B., Gary N. Atlin, Yoseph Beyene, Kassa Semagn et al. 2014. Optimal design of preliminary yield trials with genome-wide markers. *Crop Science* 54(1): 48–59. https://doi.org/10.2135/cropsci2013.03.0154.

Falconer, Douglas S. and Trudy Mackay. 2009. *Introduction to Quantitative Genetics*. 4. ed., [16. print.]. Harlow: Pearson, Prentice Hall.

Federer, Walter T. 1961. Augmented designs with one-way elimination of heterogeneity. *Biometrics* 17(3): 447. https://doi.org/10.2307/2527837.

Federer, W. T. and Raghavarao, D. 1975. On augmented designs. *Biometrics* 31(1): 29. https://doi.org/10.2307/2529707.

Federer, Water T. 2005. Augmented split block experiment design. *Agronomy Journal* 97(2): 578–86. https://doi.org/10.2134/agronj2005.0578.

Gapare, Washington, Shiming Liu, Warren Conaty, Qian-Hao Zhu et al. 2018. Historical datasets support genomic selection models for the prediction of cotton fiber quality phenotypes across multiple environments. *G3: Genes|Genomes|Genetics* 8(5): 1721–32. https://doi.org/10.1534/g3.118.200140.

Goddard, Mike. 2009. Genomic selection: prediction of accuracy and maximisation of long term response. *Genetica* 136(2): 245–57. https://doi.org/10.1007/s10709-008-9308-0.

Goiffon, Matthew, Aaron Kusmec, Lizhi Wang, Guiping Hu et al. 2017. Improving response in genomic selection with a population-based selection strategy: optimal population value selection. *Genetics* 206(3): 1675–82. https://doi.org/10.1534/genetics.116.197103.

Golicz, Agnieszka A., Jacqueline Batley and David Edwards. 2016. Towards plant pangenomics. *Plant Biotechnology Journal* 14(4): 1099–1105. https://doi.org/10.1111/pbi.12499.

Gorjanc, Gregor and John M Hickey. 2018. AlphaMate: A program for optimizing selection, maintenance of diversity and mate allocation in breeding programs. Edited by Oliver Stegle. *Bioinformatics* 34(19): 3408–11. https://doi.org/10.1093/bioinformatics/bty375.

Han, Ye, John N. Cameron, Lizhi Wang and William D. Beavis. 2017. The predicted cross value for genetic introgression of multiple alleles. *Genetics* 205(4): 1409–23. https://doi.org/10.1534/genetics.116.197095.

Hayes, B. J., Bowman, P. J., Chamberlain, A. J. and Goddard, M. E. 2009. Invited review: genomic selection in dairy cattle: progress and challenges. *J. Dairy Sci.* 92(2): 433–43. https://doi.org/10.3168/jds.2008-1646.

Hayes, B., Shepherd, R., Newman, S. and Kinghorn, B. 1998. A tactical approach to improving long term response in across breed mating plans. *In*: *Proceedings of the Sixth World Congress on Genetics Applied to Livestock Production*, 25: 431–38. Armidale, NSW.

Heslot, Nicolas, Deniz Akdemir, Mark E. Sorrells, Jean-Luc Jannink et al. 2014. Integrating environmental covariates and crop modeling into the genomic selection framework to predict genotype by environment interactions. *Theoretical and Applied Genetics* 127(2): 463–80. https://doi.org/10.1007/s00122-013-2231-5.

Howard, Réka, Alicia L. Carriquiry and William D. Beavis. 2014. Parametric and nonparametric statistical methods for genomic selection of traits with additive and epistatic genetic architectures. *G3 Genes|Genomes|Genetics* 4(6): 1027–46. https://doi.org/10.1534/g3.114.010298.

Isidro, Julio, Jean-Luc Jannink, Deniz Akdemir, Jesse Poland et al. 2015. Training set optimization under population structure in genomic selection. *Theoretical and Applied Genetics* 128(1): 145–58. https://doi.org/10.1007/s00122-014-2418-4.

Jannink, J.-L., Lorenz, A. J. and Iwata, H. 2010. Genomic selection in plant breeding: from theory to practice. *Briefings in Functional Genomics* 9(2): 166–77. https://doi.org/10.1093/bfgp/elq001.

Jarquín, Diego, José Crossa, Xavier Lacaze, Philippe Du Cheyron et al. 2014. A reaction norm model for genomic selection using high-dimensional genomic and environmental data. *Theoretical and Applied Genetics* 127(3): 595–607. https://doi.org/10.1007/s00122-013-2243-1.

Jarquin, Diego, Reka Howard, Jose Crossa, Yoseph Beyene et al. 2020. Genomic prediction enhanced sparse testing for multi-environment trials. *G3 Genes|Genomes|Genetics* 10(8): 2725–39. https://doi.org/10.1534/g3.120.401349.

Kemper, K. E., Bowman, P. J., Pryce, J. E., Hayes, B. J., Goddard, M. E. et al. 2012. Long-term selection strategies for complex traits using high-density genetic markers. *Journal of Dairy Science* 95(8): 4646–56. https://doi.org/10.3168/jds.2011-5289.

Kinghorn, B. and Shepherd, R. 1994. A tactical approach to breeding for information-rich designs. *In*: *Proceedings of the Fifth World Congress on Genetics Applied to Livestock Production*, 18: 255–61. Guelph, ON.

Lee, S. Hong, Sam Clark and Julius H. J. van der Werf. 2017. Estimation of genomic prediction accuracy from reference populations with varying degrees of relationship. Edited by Qin Zhang. *PLOS ONE* 12(12): e0189775. https://doi.org/10.1371/journal.pone.0189775.

Lehermeier, Christina, Simon Teyssèdre and Chris-Carolin Schön. 2017. Genetic gain increases by applying the usefulness criterion with improved variance prediction in selection of crosses. *Genetics* 207(4): 1651–61. https://doi.org/10.1534/genetics.117.300403.

Lin, Zibei, Junping Wang, Noel O. I. Cogan, Luke W. Pembleton, Pieter Badenhorst et al. 2017. Optimizing resource allocation in a genomic breeding program for perennial ryegrass to balance genetic gain, cost, and inbreeding. *Crop Science* 57(1): 243–52. https://doi.org/10.2135/cropsci2016.07.0577.

Lopez-Cruz, Marco, Jose Crossa, David Bonnett, Susanne Dreisigacker, Jesse Poland et al. 2015. Increased prediction accuracy in wheat breeding trials using a marker × environment interaction genomic selection model. *G3 Genes|Genomes|Genetics* 5(4): 569–82. https://doi.org/10.1534/g3.114.016097.

Lorenz, Aaron J. 2013. Resource allocation for maximizing prediction accuracy and genetic gain of genomic selection in plant breeding: a simulation experiment. *G3 Genes|Genomes|Genetics* 3(3): 481–91. https://doi.org/10.1534/g3.112.004911.

Moeinizade, Saba, Guiping Hu, Lizhi Wang, Patrick S. Schnable et al. 2019. Optimizing selection and mating in genomic selection with a look-ahead approach: an operations research framework. *G3 Genes|Genomes|Genetics* 9(7): 2123–33. https://doi.org/10.1534/g3.118.200842.

Moeinizade, Saba, Aaron Kusmec, Guiping Hu, Lizhi Wang, Patrick S. Schnable et al. 2020. Multi-trait genomic selection methods for crop improvement. *Genetics* 215(4): 931–45. https://doi.org/10.1534/genetics.120.303305.

Oakey, Helena, Arūnas Verbyla, Wayne Pitchford, Brian Cullis, Haydn Kuchel et al. 2006. Joint modeling of additive and non-additive genetic line effects in single field trials. *Theoretical and Applied Genetics* 113(5): 809–19. https://doi.org/10.1007/s00122-006-0333-z.

Roos, A. P. W. de, Hayes, B. J. and Goddard, M. E. 2009. Reliability of genomic predictions across multiple populations. *Genetics* 183(4): 1545–53. https://doi.org/10.1534/genetics.109.104935.

Santantonio, Nicholas, Sikiru Adeniyi Atanda, Yoseph Beyene, Rajeev K. Varshney, Michael Olsen et al. 2020. Strategies for effective use of genomic information in crop breeding programs Serving Africa and South Asia. *Frontiers in Plant Science* 11(March): 353. https://doi.org/10.3389/fpls.2020.00353.

Shepherd, R. and Kinghorn, B. 1998. A tactical approach to the design of crossbreeding programs. *In*: *Proceedings of the Sixth World Congress on Genetics Applied to Livestock Production*, 25: 431–38. Armidale, NSW.

Smith, A. B., Lim, P. and Cullis, B. R. 2006. The design and analysis of multi-phase plant breeding experiments. *The Journal of Agricultural Science* 144(5): 393–409. https://doi.org/10.1017/S0021859606006319.

Vallejo, Roger L., Timothy D. Leeds, Guangtu Gao, James E. Parsons, Kyle E. Martin et al. 2017. Genomic selection models double the accuracy of predicted breeding values for bacterial cold water disease resistance compared to a traditional pedigree-based model in rainbow trout aquaculture. *Genetics Selection Evolution* 49(1): 17. https://doi.org/10.1186/s12711-017-0293-6.

Verhoeven, K. J. F., Jannink, J.-L. and McIntyre, L. M. 2006. Using mating designs to uncover qtl and the genetic architecture of complex traits. *Heredity* 96(2): 139–49. https://doi.org/10.1038/sj.hdy.6800763.

Wang, Lizhi, Guodong Zhu, Will Johnson and Mriga Kher. 2018. Three new approaches to genomic selection. Edited by H.-P. Piepho. *Plant Breeding* 137(5): 673–81. https://doi.org/10.1111/pbr.12640.

Wolfe, Marnin D., Ariel W. Chan, Peter Kulakow, Ismail Rabbi, Jean-Luc Jannink et al. 2021. Genomic mating in outbred species: predicting cross usefulness with additive and total genetic covariance matrices. *Genetics* 219(3): iyab122. https://doi.org/10.1093/genetics/iyab122.

Xu, Y. 2012. Environmental assaying or e-typing as a key component for integrated plant breeding platform. *In*: *Marker-Assisted Selection Workshop*. Bento Goncalves, RS, Brazil.

Xu, Yunbi, Ping Li, Cheng Zou, Yanli Lu, Chuanxiao Xie et al. 2017. Enhancing genetic gain in the era of molecular breeding. *Journal of Experimental Botany* 68(11): 2641–66. https://doi.org/10.1093/jxb/erx135.

Zhang, Haohao, Lilin Yin, Meiyue Wang, Xiaohui Yuan, Xiaolei Liu et al. 2019. Factors affecting the accuracy of genomic selection for agricultural economic traits in maize, cattle, and pig populations. *Frontiers in Genetics* 10(March): 189. https://doi.org/10.3389/fgene.2019.00189.

Appendix I

Abbreviations

AFLP	Amplified Fragment Length Polymorphism
AM	Association Mapping
AMMI	Additive Main effect and Multiplicative Interaction
APR	Adult plant resistance
BAC	Bacterial Artificial Chromosome
BIL	Backcross Inbred Line
BMTME	Bayesian Multi-Trait Multi-Environment
BL	Bayesian LASSO
BP	Breeding population
BRR	Bayesian Ridge Regression
BV	Breeding Value
CD	Coefficient of Determination
CGS	Conventional Genomic Selection
CMS	Cytoplasmic Male Sterility
CV	Cross-validation
DArT	Diversity Array Technology
DE	Double Exponential
DH	Days to heading/Double haploid
DM	Double monoploid
EBN	Endosperm balance number
EG-BLUP	Extended Genomic Best Linear Unbiased Prediction
ELS	Early leaf spot
EM	Expectation Maximization
EN	Elastic Net
ERR	Eberhart and Russell Regression
EV	Early vigor
GAB	Genomic Assisted Breeding
GB	Genotype Building
GBLUP	Genomic Best Linear Unbiased Prediction
GBS	Genotyping by sequencing
GCA	General Combining Ability

G × E or GEI	Genotype × Environment interaction
GEBV	Genomic Estimated Breeding Value
GK	Gaussian Kernel
GM	Genomic Mating/Genetically modified
GMO	Genetically modified organisms
GP	Genomic Prediction
GPC	Grain protein content
GRM	Genomic Relationship Matrix
GS	Genomic Selection
GWAS	Genome-Wide Association Study
GY	Grain yield
HSW	Hundred seed weight
HTP	High-Throughput Phenotyping
HV	Haploid Values
IBD	Identical by descent
IBS	Identical by state
ISSR	Inter Simple Sequence Repeat
LAMS	Look Ahead Mate Selection
LASSO	Least Absolute Shrinkage and Selection Operator
LD	Linkage Disequilibrium
LLS	Late leaf spot
MAB	Marker Assisted Breeding
MABC	Marker Assisted Backcross Breeding
MAF	Minor Allele Frequency
MAGIC	Multi-parent Advanced Generation Intercross
MARS	Marker Assisted Recurrent Selection
MAS	Marker Assisted Selection
MBGRM	Multi-Breed Genomic Relationship Matrix
MCMC	Monte Carlo Markov Chain
M × E	Marker × Environment interaction
MET	Multi-Environmental trial
ML	Machine Learning
MRRS	Modified Reciprocal Recurrent Selection
MSI	Mate Selection Index
MTME	Multi-Trait Multi-Environment
MTR	Multi-Trait Ridge Regression
NAM	Nested Association Mapping
NC II	North Carolina mating design II
NGS	Next Generation Sequencing
NIL	Near Isogenic Lines
OHV	Optimum Haploid Value
OPV	Optimum Population Value

PCA	Principal Component Analysis
PCV	Predicted Cross Value
PEV	Prediction error variance
PH	Plant height
PLS	Partial Least Square
PMiGAP	Pearl Millet inbred Germplasm Association Panel
PS	Phenotypic selection
PVY	Potato Virus Y
QTL	Quantitative Trait Loci
RAPD	Random Amplified Polymorphic DNA
REML	Restricted Maximum Likelihood
RFLP	Restriction Fragment Length Polymorphism
RFR or RF	Random Forest regression
RIL	Recurrent Inbred Line
RKHS	Reproducing Kernel Hilbert Space
RR	Ridge Regression
RRBLUP	Ridge Regression Best Linear Unbiased Prediction
SCA	Specific Combining Ability
SCN	Soybean cyst nematode
SI	Selection index
SMRT	Single Molecule Real Time
SN	Seed number per plant
SNP	Single Nucleotide Polymorphism
SR	Stem rust
SSR	Simple Sequence Repeats
SSVS	Stochastic Search Variable Selection
ST- and MT-LAMS	Single-Trait and Multi-Trait Look Ahead Mate Selection
SVC	Support Vector Classification
SVM	Support Vector Machine
SVR	Support Vector Regression
TBV	True Breeding Value
TMV	Tobacco Mosaic Virus
TP	Training population
TPS	True potato seeds
TSWV	Tomato Spotted Wilt Virus
TRS	Training population set
TW	Test weight
WGR	Whole Genome Regression
WGRS	Whole Genome Re-Sequencing
WGS	Whole Genome Sequencing
YR	Yellow rust

Appendix II

Tools, platforms, and data sources

R packages and use

STPGA – training set optimization for subset selection
TSDFGS – optimization of training set by a genetic algorithm
TrainSel – training set optimization
rrBLUP – ridge regression Best Linear Unbiased Prediction
BLR – Bayesian Linear Regression
BGLR – Bayesian Generalized Linear Regression
GS3 – Genomic selection, Gibb's sampling, Gauss Seidel
Gselection – feature selection and Genomic Prediction
GenomicLand – Prediction and genomic association study
BWGS – Genomic Selection in wheat breeding
MTGS – Genomic Selection using multiple traits
BMTME – Bayesian multi-trait multi-environmental model
BGGE – Bayesian genomic genotype × environment interaction
predCrossVar – Genomic Mating

Web-based/open-source tools and use

solGS – Genomic Selection. Accessible at http://cassavabase.org/solgs.

GVCBLUP – Genomic Prediction, variance component estimation of additive and dominance effect, calculation of genomic additive and dominance relationship matrices. Accessible at https://animalgene.umn.edu/gvcblup.

GenoMatrix – pedigree based genomic prediction analysis. Available at http://compbio.ufl.edu/software/genomatrix/.

ShinyGPAS – Genomic Prediction accuracy simulator based on deterministic formulas. Available at https://chikudaisei.shinyapps.io/shinygpas/.

SeqBreed – Genomic prediction. Available at https://github.com/miguelperezenciso/SeqBreed.

GVCHAP – Genomic Prediction and component estimation using haplotypes and SNP markers. Available at https://animalgene.umn.edu/gvchap.

ISMU – Genomic Selection. Developed by ICRISAT for the internal users.

AlphaMate – genomic optimization algorithm with executable versions available in http://www.AlphaGenes.roslin.ed.ac.uk/AlphaMate.

Genomic mating code written in C++ and R available from the authors Deniz Akdemir and Julio I. Sanchez (authors of the second chapter in this book).

Data resources

https://maswheat.ucdavis.edu/, Marker Assisted Back Cross breeding projects in wheat

Pearl Millet inbred Germplasm Association Panel (PMiGAP) developed at ICRISAT in partnership with Aberystwyth University

Asian Vegetable Research and Development Center (AVRDC) gene bank in Taiwan, has a large number of germplasms for *Solanaceae* crops. https://avrdc.org/about-avrdc/history/2010s/

The United States Department of Agriculture (USDA) through Germplasm Resources Information Network (GRIN) has collected and maintaining accessions of eggplant, tomato, potato, and peppers

European Cooperative Programme for Plant Genetic Resources (ECPGR) is a collaborative program among European countries for long-term conservation and use of plant genetic resources

Triticeae Coordinated Agricultural Project (T-CAP) available at http://triticeaetoolbox.org/ is a database by USDA comprises information which facilitate Genomic Selection in dairy cattle, wheat, and barley.

Potato germplasm available at http://genebank.cipotato.org/gringlobal/search.aspx by International Potato Centre (CIP) and by USDA – ARS at https://npgsweb.ars-grin.gov/gringlobal/cropdetail.aspx?type=descriptor&id=73) and a compilation of germplasms onto a single platform is in progress at https://www.genesys-pgr.org/

Spud DB (http://solanaceae.plantbiology.msu.edu/), a potato genome database holds improved genomic resources of the International Potato Genome Sequencing Consortium (PGSC)

Potato Microarray Database at https://ics.hutton.ac.uk/solarray/

PoMaMo Database – Potato Maps and More (https://www.gabipd.org/projects/Pomamo/) hosts all the potato genetic/molecular maps, SNPs, Indels, sequence data from diploid and tetraploid genotypes developed via GABI (Genomanalyse im biologischen System Pflanze)

The Genomic Open-source Breeding informatics initiative (GOBii) (www.gobiiproject.org) and Excellence in Breeding Platform (EiB) (www.excellenceinbreeding.org) are two multi-institutional undertakings which work with public sector breeding programs in merging and storing data, facilitating cost-effective genotyping, tracking samples, implementing advanced analysis pipelines, and guiding in time-bound decision making.

Index

H

I

L

T

U

V

W

Y

Z